THE ESSENTIALS OF FACTOR ANALYSIS

to Eveline, Paul and Louise

THE ESSENTIALS OF

FACTOR ANALYSIS

DENNIS CHILD

Lecturer in the Psychology of Education
University of Bradford

HOLT, RINEHART AND WINSTON
London · New York · Sydney · Toronto

Printed in Great Britain by Billing & Sons Ltd., Guildford and London

PREFACE

This book was written with the semi-numerate in mind, in particular those students, teachers and researchers who require just sufficient knowledge of factor analysis as to make intelligible their reading of books and research papers in which the method appears. For some, I hope it will prove to be a useful starting point for a more advanced study of the subject. Experience with postgraduates reading for higher degrees in education has led me to the conclusion that the few admirable textbooks already in existence are quite difficult for the student from an 'arts' background. It is hoped this book will serve to meet the need for an elementary introduction.

Writing a technical, yet unsophisticated, text presents many problems not least of which is the risk of oversimplifying to the point of inaccuracy. Therefore, I have made every effort to keep the arguments technically acceptable whilst trying to remain sufficiently uncomplicated for the readership I have in mind. However, it is doubtful whether a complex and highly mathematical subject such as factor analysis could ever be presented in a modest volume in the form of a 'do-it-yourself' kit, and I anticipate the text will be used in conjunction with other sources of information. There is, unfortunately, no painless route to an understanding of factor analysis.

There has been an attempt, as far as possible, to avoid getting bogged down in the controversies which have occupied the minds of some factorists, and I have deliberately tried to take an unbiased position. Some references are made to controversial issues, but only a detailed knowledge of the subject would achieve the kind of insight required to cope with the *minutiae* of the arguments. For a discussion of the relative merits of various methods of analysis, the reader is recommended to look at the suggestions for further reading at the end of the book. In addition, he might find statistical journals relating to psychology and education of some value in this respect.

It is becoming increasingly difficult for students of the behavioural sciences to gain an adequate comprehension of the growth and development of important areas without some knowledge of factor analysis. In the major psychological fields of human ability and personality and to a growing extent in other sciences, factor analysis has become an inescapable and powerful technique for coping with the overwhelming number of variables operating. The general student in psychology might, therefore, find the book useful, particularly chapter 5 which shows the relationship between the technique and certain theories of human ability and personality.

I am most grateful to a number of people for their encouragement and criticisms during the writing of the book. Two of my colleagues, Louis Cohen and Derek Toomey, have, in their innocence of the subject, offered many useful suggestions for clarifying the text for the benefit of non-mathematicians. The substance of the text grew out of lectures delivered to unsuspecting advanced students with no particular head for mathematics; their questions and comments have been of tremendous value in deciding on the problems which others might experience in this field. At an early stage, Professor H. J. Butcher read two chapters, and I am grateful to him for encouraging me to complete the book. I am especially indebted to Sheila Stewart for typing the manuscript. The various quotations and tables have been acknowledged at the appropriate points in the body of the text.

University of Bradford   Dennis Child
December 1969

CONTENTS

CHAPTER ONE

ORIGINS, PURPOSES AND LIMITATIONS

A central aim of factor analysis is the 'orderly simplification', to use Burt's[1] phrase, of a number of interrelated measures. It should be reassuring for the reader to discover that factor analysis seeks to do precisely what man has been engaged in throughout history, that is to make order out of the apparent chaos of his environment. This process of identifying and classifying the attributes of our surroundings in an attempt to make our world intelligible is a very familiar one. In fact, if we were not able to organize our experiences in such a complex environment, the assimilation and communication of knowledge would be a most arduous, if not impossible, task. The process has been adequately described in the work of developmental psychologists on the formation of concepts. At an early stage of development, children gradually learn the characteristics which differentiate one object from another by observing and manipulating them in a variety of situations. This cataloguing of similarities and differences has much in common with factor analysis.

Another commonplace illustration of methods parallel to factorial analysis can be seen in the field of medicine. There are numerous ways in which ill-health becomes manifest. Temperature, heart beat, physical growths, pain and so on are observed and by noting the recurrence of certain physical symptoms the doctor builds up a cluster of symptoms which always seem to appear together—a syndrome. It is only by virtue of these regularities that medicine has been in a position to identify diseases and subsequently to find a cure. It is, as Mill postulates, the 'concomitant variation' which enables the doctor to connect particular symptoms with particular underlying causes. Of course, medical diagnosis is fraught with complications because the symptoms from one disease overlap with those from another. The combination of a headache and a high temperature, in the absence of any other symptoms, might be impossible to diagnose. Similar problems of overlap occur in factor analysis, as we shall see.

However, this example from medicine has the advantage of depending on criteria which lend themselves to fairly obvious observation and measurement. A doctor can determine body temperature or observe physical abnormalities with reasonable precision. Generalizations about the symptoms and the cure stand a good chance of being reliable and successful. When we try the same with human behaviour, our measures, such as they are, are more numerous and have greater margins of error. We must therefore recognize from the outset that conclusions have to be treated with greater caution.

The mistrust of the method probably stems from exaggerated doubts in the minds of some people that human behaviour can be accounted for with some degree of precision. This book would not have been written if the author had shared this point of view. But he is mindful of the limitations imposed by the inaccuracy of the measures and the conclusions drawn from them. As in any other scientific study, we should be looking for improvements in the precision of the measures at our disposal and not abandoning them, particularly in the absence of anything better.

One of the chief burdens of the next chapters is to define a 'factor'. However, the term crops up so frequently beforehand that a provisional definition would not be out of place. When a group of variables has, for some reason, a great deal in common a factor may be said to exist. These related variables are discovered using the technique of correlation. For example, if one took a group of people and correlated the lengths of their arms, legs and bodies one would probably find a marked relationship between all three measures. In other words, tall men would tend to have long arms and long legs and vice versa for short men. This interconnection constitutes a factor— a factor, if you like, of linear size. If eye colour or left-handedness had been correlated along with the other variables, they would not have shown any relationship and consequently would not appear in the same factor of linear size.

THE BEGINNINGS OF FACTOR ANALYSIS

It was Galton, a brilliant scientist of the 19th and early 20th centuries, who laid the foundations of factorial study. Although he did not concern himself with the kind of mathematical analysis so familiar to the subject nowadays, he nevertheless inspired two lines of thought which have been essential to the development of factorial study. The first of these was the idea that *general intellectual power* was spread in a continuous fashion from the very dull to the very bright. It was fashionable at that time to think of our behaviour as governed by mental faculties—discrete attributes such as memory and attention. These were held by the earliest psychologists to be the direct result of particular parts of the brain, hence the growth of phrenology. Moreover, a point was reached when protagonists of the faculty movement busied themselves looking for ways in which the faculties could be exercised and improved by what amounted to 'mental gymnastics'.

This idea of a common causal thread running through all intelligent behaviour was in marked contrast to the pluralistic theories expounded by faculty theorists. His idea lives on in the form of *g* the general ability formulated, as we shall discover, using factor solutions. Galton took his argument a step further by proposing the existence of *special powers*, although he still believed the general intellectual power was the overriding influence in determining the quality of a man's responses *in general.* In his book *Hereditary Genius*,[2] he optimistically declares that 'men who have no natural taste in science, and yet succeed in it, may be accredited with sufficient *general ability* to leave their mark on whatever subject it becomes their business to undertake'.

The second major contribution made by Galton was the concept of correlation. He developed quantitative methods to give some idea of the interdependence between two variables; but he did overrate the value of the correlation coefficient by assuming that it would also provide the underlying causes of the resemblance between the variables. We are now more cautious in ascribing cause or effect relationships between variables. Mill, a logician, in defining his notion of 'concomitant variation' sheds some light on the limitations present when interpreting correlation values. In effect, he says that if one 'phenomenon' varies in a similar manner to a second 'phenomenon', the latter is either a cause, an effect or there exists some common causative factor.

This point is worth pursuing because we shall be making extensive use of the concept of correlation. Take as an example the positive (low but significant) correlation which has frequently been found at secondary and higher levels of education between introversion as measured by personality questionnaires and performance in examinations. Clearly, there are many plausible explanations for this association, but the origins of the cause could not possibly be teased out from a knowledge of the correlation alone.

The student may be endowed by nature with characteristics typical of introverts. As defined by personality inventories this would mean he was not person-orientated, he was persistent, he liked dabbling with ideas and so on. Given that he has the ability to cope with the subjects in the examination, he is more likely to succeed than the extravert because this situation requires long periods of impersonal, tedious and studious activity in preparation for and performance in examinations. In this case personality has been instrumental in aiding the student's performance.

On the other hand, the qualities of an introvert might have been acquired. The school and the family might have succeeded in inculcating introvert habits for the expressed purpose of being rewarded by an examination system. The tables have been turned and the examination system becomes the source which has helped to shape the personality of the student.

Alternatively, there could be an underlying cause which accounts for both introversion and examination success appearing together. It may be that highly intelligent people tend to be introverted and, one would suspect, succeed in examinations (the evidence so far does not support this point of view). Maybe social class is the common denominator where students from middle-class homes might tend to be introverted and do well in examinations (the evidence in this respect is not convincing). The reader by now has probably raced ahead in thinking through a number of causal relationships and confirming the inadequacy of a correlation coefficient for this purpose.

Getting back to Mill's postulate for a moment, it applies equally well to a number of 'phenomena' which show simultaneous and consistent variation under similar conditions. The science of finding these related 'phenomena' is, in a nutshell, the province of the factor analyst. The giant step to inferring the underlying causal determinants is a matter for speculation, ingenuity and further research.

FIRST IN THE FIELD

Karl Pearson,[3] in a famous paper at the beginning of this century, was the first to make explicit a procedure for a factor analysis and, as we shall see in chapter 2, he derived his formulae by considering the geometry of multidimensional space. The earliest suggestion of an application for this new technique came in 1902 when Macdonnell[4] wrote a paper on the study of 'criminal anthropometry and the identification of criminals'. Using physical characteristics such as head, body and limb lengths as variables he compared 3000 criminals with 1000 Cambridge undergraduates to probe the extent to which criminals diverged in physical characteristics from the 'community' (although you will have noticed the latter was biased). Apparently, Cambridge men were taller and had bigger heads. As a matter of fact, Macdonnell suggested the difference arose from socio-economic rather than crimino-anthropometric sources.

It was Spearman's[5] report in 1904 on 'general intelligence' which heralded the intensive study of human ability using mathematical models. In this paper he posited the well-known Two-Factor Theory to be discussed in chapter 5. Elaborate and refined psychological and mathematical arguments blossomed from these early efforts of Galton, Pearson and Spearman. The names of Burt (simple summation— a forerunner of Thurstone's technique known as centroid analysis, and prolific contributor in most aspects of factor analysis), Thurstone in the 1930's (centroid analysis and the multi-factor approach), Thomson (a strong critic of Spearman's Two-Factor Theory) and a number of others whose names will appear in the text have been associated with the development of the subject.

The range of subject areas in which factor analysis has played an important role is now very extensive. One's first impression is that psychology (particularly intelligence and personality) is the only subject which has made use of the technique. Admittedly, the growth and refinement of factor methods owes a great deal to the early explorations of psychologists searching for a neat and tidy description of man's intellectual abilities. But many other applications now exist. Harman[6] mentions quite a number of researches in America including politics, sociology, economics, man–machine systems, accident research, taxonomy, biology, medicine and geology. In chapter 5, in addition to intelligence and personality studies, we shall take a look at applications from some of these other areas.

PURPOSES OF FACTOR ANALYSIS IN RELATION TO THE BEHAVIOURAL SCIENCES

All sciences endeavour to describe, predict and, in some cases, add a measure of control over the variables being manipulated.[7] Take an example from physics. The factors influencing the time of swing of a pendulum are first described as a result of accounting for all the conceivable variables. There are many factors one could select such as the weight on the end of the pendulum, the length from the point of suspension to the weight, the angle to which the weight is drawn before release, the force with which the weight is pushed when released and so on. Once the important variables have been singled out by controlled experiment, one can proceed to form generalizations about the exact relationship governing their interdependence.

In arriving at generalizations, various hypotheses have been tested; for example, what would happen if we kept all other variables constant and moved the pendulum to different places on the earth's surface? In this way, the hypothesis, and consequently the generalization, is put on trial. Generalizations (theories, laws, etc.) can then be used as a source of predictions, that is, a great many hypotheses are suggested by the results of the experiment. It might be simple as in predicting the time of swing from a given length, or it might be highly complex as when relating the phenomenon to gravitation. Only when predictions have reached a high degree of accuracy, largely as a function of the precision of the measuring instruments (animate and inanimate), is control possible and only then when man's intervention is feasible.

This train of possibilities has hardly been realized in the behavioural sciences. There are very few instances where prediction has been sufficiently accurate for the control of human activity. In fact, it has been argued by some that we are really no further than the descriptive level. Perhaps the closest we have come in psychology to the use of description, prediction and control as we saw in physics is the 11+ examination in this country. As a result of descriptive factorial studies in the twenties and thirties it was decided that a knowledge of the child's general ability, using standardized IQ tests along with tests of specific verbal and numerical skills, would be adequate for the prediction of scholastic potential. It is now common knowledge that the scores are used to predict the style of education best suited for children. In a crude sense the 'control' element of the process arises from the deliberate division of schoolchildren according to their predicted capacities. The unpopularity of the scheme arises not so much from the inaccuracy of the predictors (although these have been shown to be only partially reliable) as from the inequality of developmental opportunities for the children.

DESCRIPTIVE ANALYSIS

Whatever else we claim for factor analysis, it certainly enables us to describe a group provided, of course, that our measures are reliable. Starting with a mass of tests which show correlations we can end up with a few factors or dimensions. The factors are often taken as descriptive of the group. Such terms as 'cosmopolitan', 'introvert', or a 'primary ability' are descriptive labels given to a collection of items or tests which are highly correlated and are presumed to reflect common characteristics. It is, moreover, common practice to validate the factors against external criteria which have nothing to do with factorial methods.

I have been careful to use the expression 'group' because we tend to forget that a factor analysis is often carried out using modest and homogeneous groups. There is always the risk that the description given to a group is unique, that is, it is quite unrepresentative of any population. To overcome the problem of representation it is usual to pay considerable attention to the sampling procedure and, in addition, to employ several samples.

It would not be reasonable at this stage to give the reader an example of an experiment involving only the description of a sample because some knowledge of factor solutions is necessary. The work of Fleishman and Hempel,[8] to be mentioned in chapter 5, serves as one of the few illustrations where only the bare bones of the factor structure are used on successive occasions without recourse to the predictive nature of the factors.

A modest point of view which pinpoints the descriptive function of factor analysis is revealed in the writings of Burt[9] who thinks that

> ... we may compare the advantages of using independent factors in psychology with those of using latitude and longitude in geography . . .. Our factors, therefore, are to be thought of in the first instance as lines or terms of reference only, not as concrete psychological entities. (p. 18)

Of course, the position of the frame of reference in both geography and factor analysis is arbitrary. Whether factor analysts plump for a simple descriptive model alone or move on to modify the frame of reference in the hope of attaining greater meaningfulness is one source of controversy between them.

In most instances, the analysis is preceded by a hunch as to the factors which might emerge. In fact, it would be difficult to conceive of a manageable analysis which started in an empty-headed fashion. In selecting test material in the first place it must have occurred to the experimenter that the tests have something in common or that some are markedly different. Even the 'let's see what happens' approach is pretty sure to have a hunch at the back of it somewhere. It is this testing and the generation of hypotheses which forms the principal concern of most factor analysts.

TESTING AND CREATING HYPOTHESES

We began by presenting factor analysis as a device for ordering and simplifying correlations between related variables. However, this is usually not sufficient for the behavioural scientist. Moreover, he is faced with the dual task of simplification based on a mathematical model followed by an evaluation based on a psychological, socio-logical (or whatever) model which would add meaning relevant to his purposes. As we have argued, these purposes are precisely the same as any other science employing statistics, namely, description and inference.

Eysenck[10] is optimistic, as compared with Burt, about the possibilities for factor analysis. According to him, causal relationships *can* be inferred from the results of the procedure.

> This causal implication characterizes not only the interpretation of factors as suggestive of a hypothesis, but also the next level of factors as proving a hypothesis, and . . . from the psychological point of view this causal implication is precisely what lends interest and value to factor analysis.

His paper in the *American Psychologist* is well worth reading alongside the philosophical arguments in Burt's work (1940). The greatest danger lies in reifying the factors as if they were tangible attributes possessed in some quantity by everyone. Even so, as Burt admits, it is very difficult to avoid the language of causality and there would be little point in using factor analysis in human studies unless causal assumptions were possible. Consequently, it is important to avoid the circularity resulting from the use of factors as the only source of validation. Some external criteria are essential for substantiating factor content.

The distinction between testing and creating hypotheses in factor analysis is not very sharp. Frequently, verifying a hypothesis will throw up propositions as a by-product of unforeseen relationships. As anyone involved in research will know, results often suggest more problems than solutions. To illustrate the value of factor analysis for generating and testing hypotheses, let us return to the trivial example of a factor used earlier in this chapter. A hypothesis was created when it was assumed that arm, leg and body lengths were highly correlated and, we supposed, would represent a factor of linear size. If this had been established using factor designs, a number of additional hypotheses might have been suggested by the result. For example, we could propose and test the idea that head width or finger length belonged to this factor, or that linear size factors should be apparent in all cultures. Obviously there are many postulates which spring to mind. The verification and elaboration of hypotheses is very much at the heart of most factorial studies and several more examples will be provided in chapter 5.

An actual case of testing and setting up hypotheses is afforded by the work of Rummel[11] who has applied the technique in politics. Basically, his purpose was a systematic description of the numerous kinds of conflict which could beset a nation. Did nations experience riots, revolutions and assassinations at random or were there regular patterns of conflict? Could 'conflict dimensions' be isolated from what appeared at first sight to be a structureless domain? His first task was to define his categories of conflict which he did using detailed information from daily newspapers. He studied the domestic strifes reported in 105 nations over the period of two years from 1962 to 1964. He then grouped the conflicts under such headings as demonstrations, revolutions and guerrilla warfare. Analysis of these conflicts revealed three factors which he labelled *turmoil*, *revolution* and *subversion*. We have here a good illustration of the hypothesis testing function of factor analysis. Conflict dimensions had emerged from the mass of strifes in his classification.

The first and most prominent factor of *turmoil* involved mainly spontaneous conflicts such as riots and demonstrations. *Revolutions* embraced more systematic and open activity such as coups, revolutions and purges. The third dimension of *subversion* included the underhand activities of assassination and guerrilla warfare. The author goes on to suggest a number of hypotheses from these dimensions. For example, nations could be classified according to the magnitude of each dimension of conflict and it would be possible to predict the appearance, nature and extent of a dimension of conflict in given circumstances.

LIMITATIONS

Before we proceed with a detailed discussion of the subject, it is important to sound some notes of caution. Certain problems will have to wait until the reader has a better grasp of the topic, but there are some limitations which can be stated in general terms at this stage.

Several dangers have already been mentioned. One of the first and certainly one of the most important is to avoid reading too much into a correlation coefficient because causal relationships cannot be inferred from correlations alone. As factors are derived from correlations the same argument applies in the absence of external criteria of validation. Some test scores for example, are the end product of processes—processes such as thinking or perceiving. The scores might tell us little, if anything, about these processes, and any conclusions drawn from test analysis about the processes should be viewed with the utmost care.

It was also suggested earlier that high margins of error prevailed in tests of human behaviour. As we shall see, no entirely satisfactory method has been found in allowing for these errors. The size of the sample must enter into any speculations about errors and consequently the larger the sample the more notice we are likely to take of large correlations. The rule should be, in applying any tests of significance, to err on the side of rigour rather than leniency. Also in this connection, tests with low reliability, as we shall see later, should be avoided in factor analysis.

Sample selection is important. It often happens that a sample is homogeneous because of the special circumstances in selection of a parent population. Children chosen from a grammar school or university students selected from a technological university are highly unlikely to be typical of schoolchildren or university students in general. Samples collected from different populations should not be pooled when computing correlations. Factors which are specific to a population may become obscured when pooling is applied. Obviously, in the selection of a *sample* from a given population any of the recognized sampling procedures pertain. The problem arises when we cross from one population to another, a population being a clearly defined group of individuals. The interpretation of this definition depends on the purposes of the experiment, and this implies having a clear idea of the attributes in the population to be chosen.

The distributions of raw data will be mentioned in the next chapter. For the present, provided they are not excessively skewed, truncated or multi-modal, most kinds of distribution can be used in factor analysis. More important is the linearity of the correlation between two sets of data. Curved rather than straight-line relationships are suspect. Anything approaching a curvilinear shape should be treated with the utmost care. The scattergram in the next chapter should help to make clear the reasons for this.

Earlier in the chapter an illustration of a factor was given and called a factor for 'linear size'. This problem of naming factors has the drawback of requiring, in some cases, a notion of causal determinants. McNemar,[12] in a scathing attack on factor analysts, accuses them of suffering from the 'struggle' syndrome. 'When interpreting factors all factorists struggle and struggle and struggle' in trying to fit the factors to their initial hypotheses.

Additional limitations which apply to factor solutions will be introduced at the appropriate section in later chapters. They are largely technical difficulties and refer to erroneous ideas in the isolation, interpretation and significance of the factors.

REFERENCES

1. C. Burt, *The Factors of the Mind*, University of London Press, London, 1940. This book gives a most useful discussion of the rationale of factor analysis.
2. F. Galton, *Hereditary Genius*, Macmillan, London, 1869.
3. K. Pearson, 'On lines and planes of closest fit to systems of points in space', *Phil. Mag.*, 1901.
4. W. R. Macdonnell, 'On criminal anthropometry and the identification of criminals', *Biometrika*, **1**, 177–227 (1902). This was a biological study as the title indicates.
5. C. Spearman, 'General intelligence objectively determined and measured', *Am. J. Psychol.*, **15**, 202–293 (1904).
6. H. H. Harman, *Modern Factor Analysis*, University of Chicago Press, Chicago, 1967, p.7.
7. For a discussion of factor analysis and the theory of science see S. Henrysson, *Factor Analysis in the Behavioural Sciences*, Almqvist and Wiksell, Uppsala, 1957; C. Burt, *The Factors of the Mind*, University of London Press, London, 1940.
8. E. A. Fleishman and W. E. Hempel, 'Changes in factor structure of a complex psychomotor test as a function of practice', *Psychometrika*, **19**, 239–252 (1954).
9. C. Burt, *The Factors of the Mind*, University of London Press, London, 1940, p. 18.
10. H. J. Eysenck, 'The logical basis of factor analysis', *Am. Psychol.*, **8**, 105–114 (1953).
11. R. J. Rummel, 'A field theory of social action with applications to conflict within nations', *Gen. Syst.*, **10**, 183–196 (1965).
12. Q. McNemar, 'The factors in factoring behavior', *Psychometrika*, **16**, 353–359 (1951).

CHAPTER TWO

A GEOMETRICAL APPROACH TO FACTOR ANALYSIS

This chapter is devoted to a geometrical explanation of factor analysis. Other approaches are possible; in fact, most detailed textbooks on the subject favour an interpretation which takes the reader into the realms of matrix algebra because it enables the general case to be derived by methods familiar to mathematicians. In general, the writer has found geometrical models easier for the 'lukewarm' mathematician to follow because, at least, there is something to visualize. Conversely, matrix algebra sometimes has the effect of bewildering and discouraging the mathematically unsophisticated reader for whom this text is intended.

On the other hand, this book is not a basic text in statistics and certain assumptions have been made about the mathematical background of the reader. Those who need an elementary knowledge of factor analysis must know they cannot avoid meeting with some mathematics. In any case, they have probably had dealings with other statistical concepts. Consequently, in the oversimplified picture which follows, it is assumed that the reader has a modest acquaintance with geometry, simple algebra, trigonometry and such statistical concepts as mean, standard deviation, variance, standard scores and the Pearson product-moment correlation.

CORRELATIONS

In the last chapter, a factor was loosely defined as the outcome of discovering a group of variables having a certain characteristic in common. In order to find out if a group of variables has something in common it is necessary to know the nature of the correlations between each pair of variables. It really amounts to this, if three variables, *A*, *B* and *C* correlate with each other (*A* with *B*, *A* with *C*, *B* with *C*)* we have grounds for believing in the existence of a common relationship. Note, however, we cannot infer cause–effect relationships from this information alone. Additional criteria would be required as suggested in the previous chapter. The problem of sorting out the interrelationships becomes too complex as the number of variables increases and we have to resort to a mathematical analysis. The first task, therefore, will be to express correlations in a form suitable for geometrical analysis and from this to show how those variables with the greatest affinity (a factor) are treated.

A very useful way of portraying the correlation existing between two variables is the *scattergram*. If two tests are given to a group of people, the graph which results when each individual's scores on both tests are plotted is a scattergram. Figure 2.1 shows a hypothetical scatter of points, each point representing an individual located on the graph according to his or her score in test 1 (the vertical coordinate) and test 2 (the horizontal coordinate). There are as many points on the graph as people in the sample.

The arrangement in figure 2.1 has been idealized to give a roughly elliptical shape in which the point *A* is both the centre of the ellipse and the point of greatest density. Any person who happened to obtain the scores appropriate for point *A* would, in fact, have the sample mean score for test 1 and test 2.

The way in which the points 'swarm' can give us some idea of the sign and extent of the correlation between the test scores. The major axis of the ellipse in figure 2.1, in passing from the origin in the direction *OF*, shows a trend for scores on both tests to increase roughly in step. The individuals represented by points *B* and *C*, close to the major axis, have obtained either low scores or high scores in both tests. In other words, there is a tendency for a high positive correlation when the test scores cluster around *OF*. Individuals represented by

---

*Beware! If *A* correlates with *B* and *A* correlates with *C*, it does not follow that *B* correlates with *C*, because the underlying reason for the two relationships may be quite different.

points such as D whose scores are high on test 1 and low on test 2, or points such as E where the reverse applies, will tend to reduce the overall positive correlation. Furthermore, if the points clustered about a line joining D and E instead of OF, the overall correlation would be negative—as one variable increases the other decreases. For convenience, the diagrams have been made to represent only positive correlations.

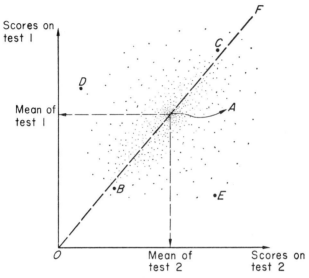

FIGURE 2.1. A hypothetical scattergram

As it stands, the 'milky-way' effect of figure 2.1 is not all that useful for estimating the extent of the correlation between the tests. On the other hand, it is possible to summarize the distributions on scattergrams by using contour lines in precisely the same manner as in geographical contours. By joining 'points' of similar density, we can get some idea of the magnitude of the correlation. The single elliptical and circular contours displayed in figure 2.2 belong to a series of concentric ellipses and circles showing the variation in density of points from the centre. In the remote case when all the points accumulate on a straight line as in (a) of figure 2.2, the correlation will be +1. The angle which the line (or the major axis of an ellipse) makes with the graph coordinates depends on the range of scores and the graph scale used. For example, with a small scale for test 1 and large one for test 2 the resulting line would be in the general direction of OA. With identical ranges and scales on both coordinates the line would be 45° to the coordinates.

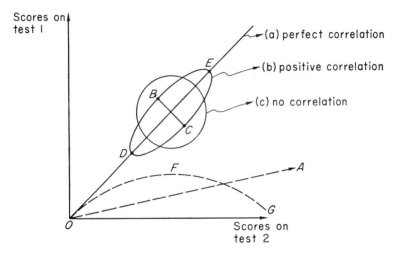

FIGURE 2.2. Idealized contours for different correlations

As the points become dispersed from the straight line into elliptical distributions [(b) in figure 2.2], the correlation coefficient could become any value between +1 and zero, the shape of the ellipse being the clue to the size of the correlation. Narrow ellipses (large major and small minor axes—DE and BC respectively in figure 2.2) show high values, but as the minor axis increases in proportion to the major axis the correlation approaches zero. When the contours are circular no correlation exists because with uniform dispersion in all directions from the centre no trends are possible.

Before pursuing the scattergram in greater detail, we need to consider, as with so many statistical arguments, the distribution of the raw test scores from which the correlations are to be derived. Fortunately, in factor analysis, most distributions are acceptable provided the arrangement of the original scores will rescale to give something approximating to a normal distribution. This will be the case in most distributions with the exception of excessively skewed, multi-modal and truncated dispersions. In fact, the original scores will often suffice without needing to transform them to a normal distribution (see Cattell[1]).

One important limitation is imposed by the relationship which exists between the test scores. Figure 2.1 was conveniently depicted as a straight-line relationship (rectilinear relationship) with the points forming a tidy cluster around the axis of the ellipse. It is possible for the resultant relationship to form a curve (curvilinear relationship). In figure 2.1 the line OF instead of being straight would bend towards one of the graph axes. Serious curvilinearity is demonstrated by the line OFG in figure 2.2. Relationships of this kind are not suitable for factor analyses because, as we shall see, factor determination is based on the assumption that correlations are derived from scores bearing linear relationships.

Up to now we have taken as read that any range of scores in the tests would suffice for comparisons between the tests. However, to assist in the description which follows, it will be convenient to convert the test scores into standard scores by dividing the difference between the mean for the sample and the individual scores by the standard deviation for the sample.* Scores lower than the mean will give negative values for standard scores. In carrying out the conversion from the mean, we have created positive and negative scores with a new unit of one standard deviation. Replotting the standard scores for both tests in the form of a scattergram would only alter the lengths of the major and minor axes of an elliptical distribution and displace the origin of the graph to the centre of the ellipse as shown in figure 2.3(b).

Note also that the origin O of the graph now occurs at the means for both sets of standard scores. Scores along OD and OA are positive and those along OC and OB are negative. Any person obtaining scores higher than the mean on both tests would be located in the first

---

*For example, if the mean for a sample is 5, the standard deviation 2 and a person's score is 7, the standard score is given by

$$\frac{7-5}{2} = +1$$

quadrant; with both scores lower than the mean, a person would be in the third quadrant. In fact, the more people we have in quadrants 1 and 3 the greater is the chance of producing a positive correlation. On the other hand, individuals with one score greater than and the other less than the mean will be found in quadrants 2 and 4 and their contribution will tend to diminish the prospect of a positive correlation.

We are now in a position to express the correlation in terms of the number of individuals occurring in the quadrants. The straight line *EF* in figure 2.3(a) clearly depicts a perfect relationship because

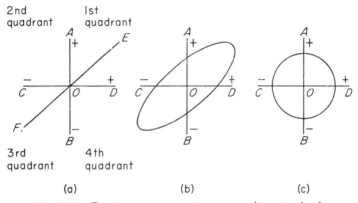

FIGURE 2.3. Contours on a scattergram using standard scores

no part of the sample appears in the disparate quadrants 2 and 4. The ellipse in figure 2.3(b) indicates larger numbers in quadrants 1 and 3, thus giving an overall positive relationship. The circular scatter in figure 2.3(c), by containing equal numbers in all the quadrants, produces complete cancellation to give a zero correlation.

THE COSINE OF AN ANGLE IS EQUAL TO THE CORRELATION COEFFICIENT

The correlation between two variables can also be expressed in terms of an angle between two straight lines. These lines, referred to as *vectors*, have special qualities because they must represent the variables in both magnitude and direction relative to each other.

The coordinates of the graphs we have used up to now could have represented the tests in magnitude provided the same scales had been used, but the graph coordinates are invariably placed at 90° to each other and therefore only represent one case. If, however, these coordinates are rotated until the *cosine* of the angle between them is numerically equal to the correlation coefficient, they become test vectors.

To illustrate the connection between a correlation and the cosine of an angle, the reader is recommended to refer to a table of cosines. In the table you will see that the cosine of 60°, for example, has a value of 0·5000. By drawing two lines of the same length (the tests have been standardized) at an angle of 60° to each other, we have expressed the correlation of 0·5000 in vector form.

Two interesting cases arise for 0° and 90°. When test vectors need to be superimposed, that is the scores correspond exactly, the cosine of the angle between them (0°) is 1·000 and, as we anticipate, the correlation is perfect. Test vectors which subtend an angle of 90°, technically known as *orthogonal*, represent zero correlation (cosine 90° = 0). It follows from the values in the cosine table that as the angle between vectors increases from 0° to 90° the corresponding correlation decreases from 1 to 0. Any angle between zero and 180°, excluding 90°, is said to be *oblique*. Obtuse angles correspond to negative correlations, so that an angle of, say, 120° gives − 0·5000.*

The concepts outlined in the previous paragraph can be illustrated using vector diagrams. *AB* and *AC* in figure 2.4 are test vectors of

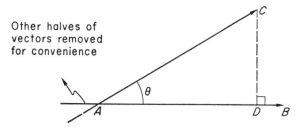

Other halves of
vectors removed
for convenience

FIGURE 2.4. Test vectors

---

*To refresh your memory of obtuse angles:
cosine 120° = −cosine (180−120)° = −cosine 60 = −0·5000

equal and unit length subtending an angle θ, the cosine of which is equal to the correlation between the tests. If a perpendicular is dropped from C onto AB (it could be from B onto AC if one wished), the cosine would be AD/AC. But if AC is of unit length, AD will be numerically the same as the correlation coefficient because the correlation being the cosine of the angle is equal to $AD/1 = AD$. Hence, if θ is 60° the length AD would be 0·5000 units, given that AB and AC are of unit length. This idea that the value of a correlation coefficient is numerically the same as the length AD, given the specifications about unit lengths, will be important to later arguments.

VECTORS AND CORRELATION MATRICES

With the introduction of a third set of test scores for the same sample, it would be necessary to account for the correlation of test 1 with 2, 1 with 3 and 2 with 3. Let us suppose that these intercorrelations were +0·5000, +0·3420 and +0·7660 respectively.* The usual way of presenting them is by means of a *correlation matrix* as illustrated in table 2.1. Notice that the values could have been repeated in the 'upper triangle', but there is no need to do this.

TABLE 2.1

A correlation matrix

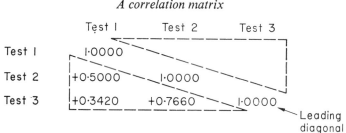

In cases where two samples have taken the same combination of tests (boys and girls, two age groups, different schools, etc.) the upper triangle may be used for one sample and the lower for the other

---

*These values have been chosen because they give convenient angles when converted to cosines. Table 2. 3 will illustrate the point.

sample. The leading diagonal (see table 2.1) of the matrix has been included to remind the reader that a test correlated with itself would give a perfect value of 1·0.

For the sake of clarity and economy, most research reports omit the positive sign and the decimal place of correlations; one should not be surprised at seeing in research literature a matrix looking something like table 2.2. The lower triangle is identical with table 2.1.

TABLE 2.2

A typical way of presenting a matrix

	1	2	3
		Girls	
1. Test 1	—	8660	1736
2. Test 2	5000	—	6428
3. Test 3	3420	7660	—
		Boys	

Decimal points omitted.

With two variables giving one correlation, all the diagrams can be presented in two dimensions, that is in the plane of the paper on which they appear. When more than two variables are involved, we come up against the problem of trying to illustrate the vectors of these variables. It is more than likely that three-dimensional space will be needed to make an accurate representation of the angles existing between the vectors.

The correlations in table 2.2 will serve as an example of how to construct the angles between more than two variables. The first task is to transpose the correlations into angles which is done using cosine tables. Table 2.3 demonstrates the values obtained from table 2.2 and can be checked from tables as an exercise.

Dealing first with the interrelations of angles in the upper triangle of table 2.3, it is possible to arrange these test vectors so that they are *all* in the plane of the page. The values have been specially chosen to give a two-dimensional display. However, this situation is extremely rare in practice. A three-dimensional array is far more likely and the example in the lower triangle of table 2.3 will reveal

TABLE 2.3

A matrix of angles derived from the correlations in table 2.2

	1	2	3
		Girls	
1. Test 1	—	30°	80°
2. Test 2	60°	—	50°
3. Test 3	70°	40°	—
	Boys		

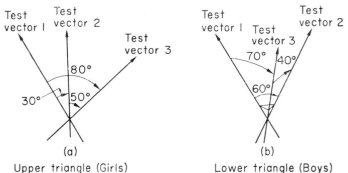

(a)
Upper triangle (Girls)

(b)
Lower triangle (Boys)

FIGURE 2.5. Arrangement of test vectors (only half the vector shown)

this. Figure 2.5 demonstrates the geometrical arrays of the test vectors for our two examples in the upper and the lower triangles of the matrix in table 2.3.

The arrangement for the vectors in figure 2.5(a) should be self-evident. The 30° between 1 and 2 plus the 50° between 2 and 3 gives 80° between 1 and 3. The configuration in figure 2.5(b) is more difficult to imagine because we have to juggle with the three vectors in space until they possess the correct inter-angular relations. If you experiment with the angles you will soon discover that it is quite impossible to have all three vectors in the same plane. Instead, imagine vectors 1 and 2 as in the plane of the page and vector 3 as coming out of the page at an angle of 70° with vector 1 and 40° with vector 2. The small triangle near the apex of the vectors in figure 2.5(b) is intended to give a three-dimensional effect.

With more than three tests the model becomes complicated, particularly when the drawings we are obliged to use are in two dimensions. Moreover, most actual analyses require more than three dimensions in order to accommodate all the angular combinations; multidimensional space, which is beyond the imagination, is a mathematical possibility referred to as *hyperspace*, but in the interests of clarity it will be safer to avoid hyperspace models. Thomson's book,[2] which illustrates the geometry of factor analysis, gives a very useful example of a three-dimensional analysis using vector diagrams.

VECTOR RESOLUTION

Given a large number of test vectors, is there any economic way of expressing the complex of associations with mathematical precision? The centroid method, to be described in the next section, was one of the earliest and neatest in the development of factor analysis. It is based on the introduction of another vector which acts as a reference line from which the test vectors are interpreted. A useful analogy is a half-open umbrella where the radiating frame gives the directions of the test vectors and the handle is used as a reference vector. These lines of reference needed for the resolution of test vectors are referred to as *common factor vectors* or simply *factors*.

To illustrate the process of finding factors by resolving vectors, a process known as *extracting factors*, let us first look at a simple vector resolution using a symmetrical two-dimensional arrangement. In figure 2.6 there are two vectors of equal length OA and OB subtending an angle of 60° (which might have represented a correlation of +0·5000 in a factor analysis). For ease of description only one half of each vector will be used in subsequent diagrams. The resultant vector 1 occupies a position along OC between the vectors and this would represent the first factor vector in an analysis (hence the 'centroid' method). Since the vectors are equal in length, the resultant vector will bisect the angle between them. Thus, in figure 2.6, OC cuts the angle between OA and OB (i.e. 60°) into two.

In order to describe the vectors OA and OB in terms of the resultant OC, we determine the angles subtended between OA and OC (30°), OB and OC (30°) then find the cosines of these angles. We are now using the resultant as a frame of reference. This cosine of the angle subtended between a test vector and a factor vector is termed a *loading* or *saturation* in factor analysis. By employing the cosine of the angle, we have, quite deliberately, expressed the loading as a correlation between the test variable and the factor vector. In figure 2.6 the loading will be cosine 30° = +0·8660 for both vectors. However, this is not the whole story because using a single line of reference does not completely define the position of the vectors in two-dimensional space. A second reference line, OD, orthogonal

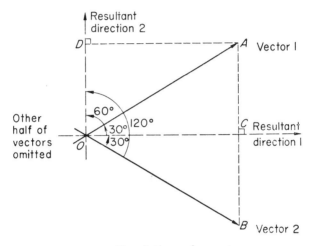

FIGURE 2.6. Resolution using vectors

(at 90°) to the first, should be introduced to complete the spatial requirement. For factorial purposes, the loadings for test vectors could readily be found from the geometry of the arrangement. The angle between OD and OA is 60° giving a loading of cos 60 = +0·5000; the angle between OD and OB is 120° which gives a loading of cos 120 = −cos 60 = −0·5000. Had this been a factor problem we would have expressed the results as set out in table 2.4.

For any one vector completely resolved, if the loadings are squared and these squares added, a total of *one* is obtained. In our example, for the first vector the loadings squared and added will give

$$(0·8660)^2 + (0·5000)^2 = 0·7500 + 0·2500 = 1·0000$$

TABLE 2.4

Factor loadings for test vectors in figure 2.6

	Loadings on first common factor vector (resultant 1)	Loadings on second common factor vector (resultant 2)
Test vector 1	0·8660	0·5000
Test vector 2	0·8660	−0·5000

For the second vector we obtain*

$$(0·8660)^2 + (−0·5000)^2 = 0·7500 + 0·2500 = 1·0000$$

The proof requires some simple trigonometry and Pythagoras. For the benefit of conscientious readers, a proof is given in the following paragraph.

To calculate the first loading in figure 2.6 for the first vector we used the triangle OCA and found the cosine of angle $COA = OC/OA$. The second loading was derived from the triangle ODA where cosine DOA was employed $= OD/OA$. Squaring and adding these two cosines gives

$$\frac{OC^2}{OA^2} + \frac{OD^2}{OA^2} = \frac{OC^2 + OD^2}{OA^2} \quad (1)$$

Applying Pythagoras' Theorem in triangle OCA,

$$OA^2 = OC^2 + AC^2 = OC^2 + OD^2 \text{ because } AC = OD$$

Substituting OA^2 in equation (1) for $OC^2 + OD^2$, we obtain

$$\frac{OA^2}{OA^2} = 1$$

Notice that the first resultant having been placed in the 'best' position from the outset has extracted higher loadings than the second resultant.

---

*The reader will recall from schooldays that in squaring a minus quantity, a 'minus times a minus gives a plus' quantity. For the second vector, $−0·5 × −0·5 = +0·25$.

THE GEOMETRY OF THE CENTROID (OR SIMPLE SUMMATION) METHOD OF FACTOR
ANALYSIS (SEE APPENDIX A FOR A MATHEMATICAL PROOF)

Suppose we had several test vectors to contend with and we wanted to
find the factor vectors and loadings for these. As in the previous
illustration, the correlations between the tests are first calculated and
the values used to find the angles between the test vectors. The
resultant arrangement of vectors is then used to find the factor vectors.

	T_1	T_2	T_3	T_4	T_5
T_1	—	10	70	90	100
T_2	9848	—	60	80	90
T_3	3420	5000	—	20	30
T_4	0000	1736	9397	—	10
T_5	−1736	0000	8660	9848	—

Upper triangle gives angles between vectors, lower gives corresponding
correlations.

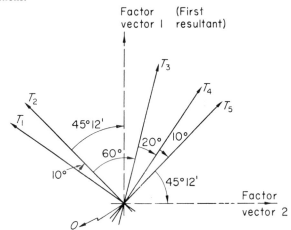

T = test vector

FIGURE 2.7. Vector diagram and matrix

As an illustration, imagine, as in figure 2.7, the vectors of five
tests to be related in such a way as to appear in the plane of this
page when all the vectors are drawn with the correct angles between
them. In normal circumstances this could be an arduous task because

all the vectors have to be manipulated until the cosines of the angles between them correspond to the correlations. Consequently, to make the illustration easier to follow, angles have been selected so that the vectors when arranged are all in the plane of the paper. The chances of this happening in an actual problem are remote. Also for the sake of clarity, whole-number angles have been used.

The first task is to locate the resultant of all five tests (the centroid) to create the first factor. The resultant is normally determined by a calculation which should not concern us here, but some idea of the concept can be gained if you visualize a stationary object at O in figure 2.7 to which five ropes (representing tests T_1 to T_5) have been tethered. If five men pull on the ropes with equal force in the directions shown, the object will slide in the direction of the resultant. The resultant in this instance is the factor vector 1. The next task is to express the position of the test vectors T_1 to T_5 relative to the factor vector 1. Thus T_1 subtends an angle of 55° 12′ to the first factor vector. Similarly, T_2 subtends 45° 12′, T_3 subtends 14° 48′, T_4 subtends 34° 48′ and T_5 subtends 44° 48′, all with the first factor vector. The cosines of these angles now give us the correlation between test vectors and the factor vector. Table 2.5 illustrates this under column I in 'Factor loadings'. The reader will discover by reference to cosine

TABLE 2.5

Loadings of test vectors in figure 2.7

	Factor loadings		Sum of squares of loadings for each test
	I	II	
1. Test 1	5707	−8211	1·0000
2. Test 2	7046	−7096	1·0000
3. Test 3	9668	2554	1·0000
4. Test 4	8211	5707	1·0000
5. Test 5	7096	7046	1·0000
sum of squares of loadings for each factor	2·9347	2·0653	= 5·0000

Decimal points omitted.

tables that the first angle for T_1 of $55°$ $12'$ gives a cosine of $+0.5707$. Other values in this first column are discovered in the same manner.

A second factor vector is now erected at right-angles (orthogonal) to the first, and calculations of the same nature carried out for the angles between T_1 to T_5 and the factor vector 2. These are shown in column II of table 2.5.

As in the correlation matrix, decimal places and positive signs have been omitted. Note once more that the squares of the loadings for any particular test will add up to one. Thus Test 1 would be

$$(0.5707)^2 + (-0.8211)^2 = 0.3257 + 0.6743 = 1.0000$$

The sum of the squares of loadings for the factors is shown at the bottom of each factor and these give, by symmetry, precisely the same total as the last column in the table (columns read down, rows read across). Ideally, the total sum of squares of the loadings will equal the number of tests, in this case five. We shall see in the next chapter that in practice the sum of squares of loadings for a test never reaches unity. Note also as with the previous example, the sum of squares for the first factor is greater than for the second.

There is nothing particularly sacred or binding about the position of the factor vectors used as reference coordinates, since any position would give a set of loadings which could be used with equal confidence. Clearly, there are advantages in adopting settings of the factor vectors that are readily reproduced because of the geometrical properties of the array. In the example we have just worked through, the reference axis, that is the factor vector, was taken to be the position occupied by the resultant of the test vectors. In a sense we are trying to locate the position of the 'handle' of the umbrella analogy used earlier and most alternatives to the centroid method give very similar positions for the factor vector and hence values of the loadings.

One very common method nowadays is called 'Principal Components' analysis. It is not intended to give any detail here, but the method utilizes the major and minor axes of the elliptical distributions as the datum lines for factor loadings. Reference to figure 2.2 shows that the factor vectors would be the major axis DE and the minor axis BC. Three-dimensional systems would have a major and two minor axes and the ellipsoidal shape generated would look like a rugby football.

Obviously, no factor analyst would ever dream of using such a long-winded process as this geometrical way of locating factors, particularly with hyperspace problems requiring multidimensional analysis of contorted and unimaginable shapes. Nowadays, computers have opened the way for rapid algebraic calculations to be performed in estimating the position of 'centroids' or any other measure for the resultant of vectors. Matrix algebra[3] is now the vehicle by which estimates are made of factor positions amongst test vectors. Behavioural scientists have grown to rely on mathematicians for finding accurate and efficient equations suitable in factor solutions.

Several important concepts have been introduced in this chapter through the medium of geometrical models. We shall need most of the concepts in the development which follows, and it will be taken for granted that such terms as factor, vector, loading, orthogonal, oblique and correlation matrices have become established in the reader's repertoire!

REFERENCES

1. R. B. Cattell, *Factor Analysis*, Harper, New York, 1952, p. 328.
2. G. H. Thomson, *The Geometry of Mental Measurement*, University of London Press, London, 1954.
3. The reader might like to see a simple presentation of factor solutions using matrix algebra. A little-known book in this country: C. J. Adcock, *Factor Analysis for Non-mathematicians*, Melbourne University Press, Melbourne, 1955, provides examples which some might find of interest.

CHAPTER THREE

THE LANGUAGE AND INTERPRETATION OF FACTOR ANALYSIS

So far we have examined a procedure for establishing frames of reference (factors) in a group of test vectors so as to create a criterion for interpreting the interrelatedness of the tests. The next stage is to expand these general principles and relate them to typical data which the reader will find in the research literature of the behavioural sciences.

VARIANCE

In the discussion of test* scores being represented by vector diagrams, we decided to simplify matters by converting the raw scores to standard scores, followed by making the full range of these standard scores equal to one. This has the effect of making any length along the vector a proportion of one, and comparable, as was seen in figure 2.4, with correlation coefficients.

To illustrate the point, suppose an IQ test has a population mean of 100 and a standard deviation of 12 IQ points. When the range on either side of the mean is converted to standard scores (sometimes referred to as z-scores), the mean would be made equal to zero and the unit would be one standard deviation; as we make 12 inches equal to 1 foot, so we make 12 IQ units equal to one unit of standard deviation.

---

*'Test' has been used in preference to the more popular term 'variable' to avoid confusion with the word 'variance' which appears frequently in this section.

The second modification makes the total range equal to one, and in figure 2.4 this was represented by the vectors AB and AC. Pursuing the IQ illustration in the previous paragraph, the range, let us say, extends from 64 to 136 IQ points. In other words, from the mean to either limit of the range (from 100 to either 64 or 136) represents three standard deviation units. We have made this equal to one unit— rather like making 3 feet equal to 1 yard. All the measurements along the test vector now appear as proportions of one. In the length analogy, all measurements would be converted to a proportion of 1 yard.

Turning again to figure 2.4, it was shown that the length of AD, provided AB and AC are unity, would be numerically equal to the correlation between the vectors (or a loading if AB is a factor vector) and, from the arguments in the previous paragraph, would be a measure of the standard deviation.

Many of the fundamental ideas in factor analysis derive from the concept of *variance* and the next step is to relate the above deductions to this concept. Variance is a very common statistical term which provides an index of the dispersion of scores. In fact, as readers will recall, it is the square of the standard deviation. Thus if we square the length AD in figure 2.4 we obtain the proportion of the variance of the test represented by vector AB which is common with the test represented by vector AC. A perpendicular from B to AC, by the symmetry of the arrangement, would have given us precisely the same numerical value along the vector AC.

By the same token, since the length AD also represents the value of the loading, the *variance is equal to the square of the loading*. If the derivations in the foregoing are forgotten, it is important to remember this final conclusion. Using a figure from table 2.5 as an illustration, the variance extracted in the first factor for the first test will be $(0 \cdot 5707)^2 = 0 \cdot 3257$.

What is the total variance of a test? There are two important components of variance required to account for the total variance of a test. These are *common* and *unique* variance.

When a factor contains two or more tests with significant* load-

---

*'Significant' in this context is used as a statistical term and refers to the size a loading must attain before we have confidence that it exists at all! More will be said later in the chapter about this, but for the moment we can take $+0 \cdot 30$ as the minimum value for a loading.

ings (or variance if the values are squared) it is referred to as a *common factor* and the variance of the tests in that factor is known as *common variance*. Since common factors account for the intercorrelations between variables there are bound to be at least two tests involved. We are back to Mill's 'concomitant variation', that is the simultaneous rise and fall in test scores enabling us to distinguish those tests with common properties. In table 2.5, as all the loadings in the first factor are high, it is a common factor. As it happens the second factor is also common because tests 1, 2, 4 and 5 have high loadings (test 3 loading is less than 0·30).

The primary aim of factor analysis is the discovery of these common factors. The techniques for extracting the factors generally endeavour to take out as much common variance as possible in the first factor. Subsequent factors are, in turn, intended to account for the maximum amount of the remaining common variance until, hopefully, no common variance remains. As we shall see in chapter 4, there are also ways in which the common variance can be redistributed to give alternative spreads of variance amongst the common factors.

Common factors also come in two sizes which readers familiar with the psychology of human ability will have met. *General* factors, usually the first in a factor solution giving the maximum variance in the first factor, include significant loadings from most if not all the tests in the analysis. *Group* factors, as the term implies, arise when a few tests with significant loadings appear in the same factor. Often several group factors occur in the same analysis.

There remains that part of the total variance of a test resulting from the unique properties possessed by the test and as such would be entirely uncorrelated with the other tests in a particular analysis. This is referred to as *unique variance*. A factor containing only one significant loading for a particular test would be a *unique factor*. The absence of any other significant loadings in the factor demonstrates a factor vector which is close to the significant test vector but orthogonal to all other tests.

Unique variance can be broken down into two further elements of *specific* and *error* variance. Each test possesses some particular qualities which are not shared with any other test in the battery under consideration, and the variation in scores arising from these qualities will produce specific variance. Of course, there may be other tests not used in a particular analysis which do in fact share some of these qualities. If they were introduced into the analysis, that portion of the specificity would now become common variance. It is, therefore, possible to alter the factor structure by introducing or eliminating tests.

Error variance, or *unreliability* as it is sometimes called, results from the imperfections of test measurement. The difference between this and the total test variance does give a measure of the reliability of the test. Unfortunately, factor analysis does not discriminate between specific and error variance, so we cannot put this knowledge to use.

As a matter of interest, on turning back to figure 2.6 and table 2.5 no unique variance is present because the model was purposely constructed to account for all the variance of the tests in two common factors. Had we wanted to complicate matters, two further dimensions (which could not be imagined anyway) might have been added to represent unique variance for each test. They would have to be orthogonal to each other and to the common factor vectors because there is no correlation between them.

In summary, we have the total variance (V_T) of a test made up from common variance (V_C) and unique variance (V_U). As the variance is additive, the relationship can be expressed in its simplest form as

$$V_T \quad = \quad V_C \quad + \quad V_U \qquad (1)$$

| total variance of a test | common variance | unique variance |

If the unique portion is divided into specific (V_S) and error (V_E) variance we have

$$V_T = V_C + V_S + V_E \qquad (2)$$

Now the common variance, it will be remembered, may be spread out amongst several factors. Strictly speaking, V_C consists of $V_{C_1} + V_{C_2} + V_{C_3} + \ldots + V_{C_n}$ where each symbol V_{C_1}, and so forth, represents the variance in each factor up to the last common factor labelled V_{C_n} (where n is the total number of common factors). It is conceivable for the total number of common factors to equal the number of tests if each occupied a plane of its own. In this case there would be n tests giving n factors. Furthermore, the unique variance would occupy yet another set of n planes giving a total of $2n$ factors. But this number is never extracted.

If the total variance of the test was made equal to one, the contributory variances on the right-hand side of the above equations would become proportions of the total variance.* Let these proportions be $C_1, C_2, C_3, \ldots, C_n$ (where n is the total number of common factors) for the common variance, S for the specific variance and E for the error variance. Equation (2) becomes

$$1 = C_1 + C_2 + C_3 + \ldots + C_n + \quad S \quad + \quad E \quad (3)$$

| total variance made equal to one | proportions of common factor variance | proportion of specific variance | proportion or error variance |

This important fundamental statement is known as the *factor equation*. An equation of this kind applies to each individual test in a factor analysis. On taking the square roots of the variances we would obtain the loadings assigned to the factors. Hence, $\sqrt{C_1}$ would be the loading of the test in factor 1, and similarly for the other variances in the remaining factors.

COMMUNALITY

The sum of all the common factor variance of a test is known as the *communality* (h^2), that is the variance shared in common with other tests. In equation (1) the communality would be V_C, whilst in equation (3) it would be the sum of $C_1 + C_2 + C_3 + \ldots + C_n$. By bringing

---

*For those interested in the mathematics of this statement, if the equation (2) is divided through by V_T, we obtain

$$\frac{V_T}{V_T} = \frac{V_C}{V_T} + \frac{V_S}{V_T} + \frac{V_E}{V_T} = 1$$

The quantities V_C/V_T, and so on, express the common variance as a proportion of the total variance.

$S + E$ to the left-hand side of equation (3), another expression is possible for h^2:

$$1 - (S + E) = h^2 \qquad (4)$$

The communality will also be *the sum of the squares of common factor loadings for a test.*

The reliability (r) of a test, as we have seen, is the sum of all the non-error variance, that is $V_C + V_S$. If we knew the reliability coefficient of a test worked out from other criteria, it would be possible by manipulating equation (3) to find the error and specific variance.* The reliability can also be used to give an indication of the thoroughness of factorization. Expressed as a percentage, the formula is

$$\frac{\text{communality}}{\text{reliability}} \times 100$$

FACTOR AND COMPONENT ANALYSIS

There are two basic models which can be adopted in factor solutions. They are known as the *factor analysis* and the *component analysis* models. Without becoming too technical, the distinction is that in factor analysis some account is taken of the presence of unique variance whereas in component analysis the intrusion of unique variance is ignored. Equation (1) above is typical of the factor analysis model. In a component analysis the unique variance becomes merged with the common variance to give hybrid 'common' factors containing small proportions of unique variance; but not enough in the

---

*Using h^2 to represent the sum of C_1 to C_n, we have

$$1 = h^2 + S + E \quad \text{from equation (3)}$$

but

$$r = h^2 + S$$

of which r and h^2 are known, thus

$$S = r - h^2$$

which gives us the specific variance. If we now replace $h^2 + S$ by r in the first expression, we have

$$1 = r + E$$

and

$$E = 1 - r$$

first few important factors, according to some authorities, for us to be worried about the overall picture obtained from the analysis.

The discrimination between these models, therefore, rests entirely on the assumptions one makes about the portions of the unit variances of each test which are to appear in the common factors. This is determined by the figure placed in the diagonal of the correlation matrix (see table 2.1), because this diagonal value defines the total common variance (communality) of a test to be distributed amongst the common factors.

To illustrate this point, turning back to table 2.1 will reveal that we assumed the leading diagonal values to be unity. If we use this value, we are saying, in effect, that the test is completely reliable and without error. The same goes for figures 2.6 and 2.7 although we did not insert unities. Nevertheless, by squaring the loadings to give us the variance, we ended up with communalities equal to one because we started off by assuming *all* the variance was common. This is the component analysis model. Within the loadings obtained there may be unique variance, but we are not to know.

If we had some knowledge of the common variance of a test before commencing a factor solution and inserted this communality in the appropriate leading diagonal for the test we would automatically build into the model an allowance for unique variance. It is all the variance which remains after our predetermined communality has been accounted for $(1 - h^2)$. This is the factor analysis model. One of the dilemmas in this form of solution was finding an efficient and accurate procedure for determining the value of the communality *before* the analysis began! The dawn of computerized solutions has made an immense difference to the methods now possible for approximating the communality.* It is not important at this stage to know how the communality is derived by the computer, but simply to know that good approximations can be realized. One of the earliest methods which is still in use is to take the largest correlation coefficient for

---

*You may have come across the expression 'iterative' or 'iteration' with regard to finding communalities. All this means is that starting from an informed guess as to the possible value of a communality, a more reliable value is calculated by repeated approximations until the final value alters very little with repeated calculations.

a row of a correlation matrix and insert the value in the leading diagonal for that row. For this, one needs the complete correlation matrix for the upper and lower triangles similar to table A.1 in Appendix A. To illustrate the point from this table, on looking along the first row representing the correlations for T_1, the largest value is 0·9848 and this would be entered as the communality instead of 1·0000. The method is applied in the example given in chapter 4, table 4.1.

A FACTOR ANALYSIS

The outline above is perhaps best summarized with an illustration from an actual factor analysis. As we have indicated above, the fundamental concern is the extraction of common factors and the discussion will not pay too much attention to unique variance.

In a recent study endeavouring to discover the cognitive and affective styles of arts and science specialists, the author presented a variety of tests to a sample of 306 university students. These included measures of verbal and spatial intelligence and several divergent thinking tests* which will be pulled out from the study to demonstrate some of the general principles outlined earlier and for use later in the text. A great deal of discussion and controversy has surrounded the subject of convergent and divergent thinking.[1] Some have suggested that for certain purposes these two modes of thinking can be treated as independent qualities. The particular results of the present analysis are not intended to add anything to the resolution of the dilemmas which exist in this field. In fact, the number of IQ tests (two measuring different aspects of general ability) is insufficient for any firm conclusions about factor content. To establish a factor, it is desirable to have at least three tests purporting to represent a dimension. However, these data have the advantage for our purposes of containing a small number of variables and factors.

---

*Divergent thinking tests are said to give respondents an opportunity to display fluency, variety and originality in solving problems which have many equally acceptable solutions, for example, 'how many uses can you think of for a piece of rope?' In contrast, a convergent test item has only one solution clearly obtainable from the information provided. Intelligence tests are said to require convergent problem-solving skills.

The IQ test used was the AH5 test[2] of high-grade intelligence which gives separate scores for verbal and spatial intelligence and a total score from the sum of the verbal and spatial elements. The latter was not used in the factor analysis because a variable should not be created from the manipulation (addition, subtraction and so forth) of variables already in the analysis. Obviously, there will be a correlation between the derived variable and those from which it was derived. This amounts to artificially creating a factor which, unless the researcher is quite clear about the implications, is not a wise course of action.

Three divergent thinking tests were taken from American researches[3] and adapted for use in this country.[4] They are referred to as 'Uses of objects', 'Consequences' and the 'Circles' tests. The first two depend upon verbal responses whilst the Circles test requires the respondent to produce drawings of familiar objects using the circle as the starting point—a non-verbal response. Two kinds of score were obtained for each test, namely, fluency and originality.[5]

We have, therefore, eight scores for each person in the sample. These scores were intercorrelated using Pearson's product-moment method. This is the recommended method for factor analysis.[6] Table 3.1 is the correlation matrix which ensued.

TABLE 3.1

Correlation matrix of IQ and divergent tests $(N = 306)$

Abbreviated title of test	1	2	3	4	5	6	7	8
1. AH5 verbal								
2. AH5 spatial	54							
3. Uses—F	08	01						
4. Uses—O	18	05	58					
5. Consequences—F	20	07	51	46				
6. Consequences—O	13	–01	26	40	46			
7. Circles—F	10	08	46	27	40	11		
8. Circles—O	05	–00	22	22	21	18	51	

Decimal points omitted.
F = Fluency; O = Originality.

Usually, to disentangle, by inspection, the complex of inter-relations in a matrix such as this would be a tedious and mathematically crude procedure. Nevertheless, we can gain some useful initial information from a consideration of the highest correlation coefficients in the matrix. One convenient way of approaching the matrix is to take the columns one by one looking for the highest coefficients in each. Another possibility is to look for blocks of significant correlations which, for a sample of 306, could be values exceeding ± 0·15. This figure was arrived at by interpolating from the table in Appendix B which readers may find helpful if they ever need to decide on the significance of correlation coefficients. The table shows the size a value must reach to be judged significant for different sample sizes. Coefficients for a sample of 300 (the nearest round figure to our sample of 306) would need to be at least ±0·15 at the one per cent level and ±0·11 at the five per cent level. Sample size is clearly an important consideration in deciding the level a correlation must attain to be significant. With small samples (30 or less) the correlation coefficients are quite unstable. The addition or omission of two or three individuals' scores can make a noticeable difference to the correlation value.

As it happens, the highest correlations do tend to form into conspicuous patterns which have been highlighted by using bold type. The correlation of 0·54 between the verbal and spatial scores is particularly interesting because on looking down the columns 1 and 2 there is not another coefficient approaching this value. Tentatively, we might conclude that the AH5 verbal and spatial scores have a lot more in common with each other than with the divergent thinking tests; note in this respect that column 2 does not contain a single significant value. Do not be put off by the appearance of a correlation coefficient of −0·00 as occurs between variables 2 and 8. In the first place, the correlations were calculated to four decimal places giving −0·0025; to two places this gives −0·00.

The values (in bold) of all the correlations for the divergent thinking tests are significant at the one per cent level except between variables 6 and 7 which qualifies at the five per cent level. This appears to be a second cluster of related variables. Our eyes could wander round the matrix picking up other minor threads of relationships, but this is precisely the function of a factor analysis. In addition to saving us the trouble of scouting through the table, it will add some degree of precision to the extent and comparability of the different relationships.

The correlation matrix now becomes the starting point for a direct analysis which in this case was a Principal Components method* using unities in the leading diagonals to represent the communality. The programme adopted was devised by Hallworth and Brebner and is available from the University of Birmingham.[7] The factor matrix obtained is displayed in table 3.2 in a form similar to that received from the computer.

TABLE 3.2

Principal components matrix for IQ and divergent tests

Tests	Common factor loadings			Communality h^2
	I	II	III	
1. AH5 verbal	0·3233	0·8150	−0·0103	0·7688
2. AH5 spatial	0·1696	0·8574	−0·1521	0·7870
3. Uses—F	0·7610	−0·1807	0·0714	0·6169
4. Uses—O	0·7349	−0·0511	0·3046	0·6354
5. Consequences—F	0·7694	−0·0199	0·2185	0·6402
6. Consequences—O	0·5658	−0·0542	0·4871	0·5603
7. Circles—F	0·6582	−0·1373	−0·5746	0·7824
8. Circles—O	0·5072	−0·1922	−0·6202	0·6788
Latent root (extracted variance, eigenvalues or sum of squares)	2·8551	1·4937	1·1210	5·4698
Percentage variance	35·6882	18·6714	14·0122	68·3718

F = Fluency; O = Originality.

First compare this table with table 2.5. The communality (h^2), that is the sum of the squares of the loadings, does not reach unity for any test because we have only been concerned with extracting common variance and our communalities must of necessity be less than one.

*The principal components are the major and minor axes of the ellipsoidal distributions alluded to in the previous chapter.

The factor loadings for test 1 when squared and added give

$$(0.3233)^2 + (0.8150)^2 + (-0.0103)^2 = 0.7688$$

The unique variance from equation (4) will be

$$1 - 0.7688 = 0.2312$$

although there is little we can do with this information unless we know the reliability of the test. Nevertheless, if the communality is too low, say in the region of 0·3 or less which gives a unique variance of 0·7 or more, it could well mean that the test is unreliable. As we saw earlier, the unique variance is the sum of specific *and* error variance or unreliability and the latter could be making the major contribution. One would be justified in eliminating that test in a re-analysis.

The sum of squares of the loadings at the bottom of each factor is technically known as the *latent root* (or alternatives such as *eigenvalue, extracted variance* and *sum of squares*). Latent root and eigenvalue are expressions taken from matrix algebra. Note their values fall off from the first factor because, as suggested earlier, the factor procedure extracts the maximum possible variance for each factor in turn. Also notice the sum of the latent roots (2·8551 + 1·4937 + 1·1210 = 5·4698), by the arithmetical symmetry of the arrangement, is identical to the sum of the communalities.

Ideally, for eight tests each with a total variance of unity the maximum possible variance is eight. The actual variance extracted in the first factor is 2·8551. A convenient way of expressing the latent roots is to convert them to percentages of the maximum possible variance. In the first factor the *percentage variance* is

$$\frac{\text{Latent root}}{\text{Number of tests}} \times 100 = \frac{2.8551}{8} \times 100 = 35.69\%$$

Percentage variances calculated for the other factors are shown in table 3.2. This figure gives us some idea of the contribution of each factor to the total variance. The higher this figure is, the more substantial can be the claim that the items with significant loadings have some property in common. The latent root of factors can also be used as a possible criterion for recognizing when all the common factors have been extracted, as we shall see. Note also the total extracted variance, which in table 3.2 was 5·4698 or 68·37 per cent of the maximum of 8, gives an idea of the variance attributable to interrelatedness.

When we come to explore the table in detail, interest will be centred on those loadings with high values. On looking at the last of the three factors the highest values are negative. All it means is that the computer has 'read the angles' between the test vectors and the factor vector with the latter at 180° to the usual position. By reference to figure 2.6, if resultant vector 1 had been placed pointing in exactly the opposite direction negative values would have been obtained. Consequently nothing is violated by reversing *all* the signs in any *one* factor if we so desire. To make the largest numbers positive, the signs have been reversed in factor III in the revised table 3.3 later in the chapter. There is no need to reverse the signs in the first or second factors, for the highest values are already positive.

CRITERIA FOR THE NUMBER OF FACTORS TO BE EXTRACTED

But how do we decide on how many factors to extract? Why was the extraction process stopped in the present example at the third factor? Essentially, only the common factors are required and the methods employed rest upon assumptions as to when this has been achieved. There are minor differences of opinion between factorists as to the best grounds for halting extraction. They have been guided chiefly by experience in adopting particular criteria, although some methods depend for their justification on mathematical interpretation. The following two popular methods are offered simply to give the reader an idea of criteria in use. There are several more and readers are advised to consult advanced texts[8] for reasoned arguments about the wide variety of methods available.

(1) A technique in considerable use at present is *Kaiser's criterion* suggested by Guttman and adapted by Kaiser. The rule is very simple to apply. Only the *factors having latent roots greater than one* are considered as common factors. This method is particularly suitable for principal components designs and it was adopted for the above example in table 3.2 where the first three factors all have latent roots greater than unity—the third factor just exceeds unity (1·1210). Cattell has suggested that Kaiser's criterion is probably most reliable when the number of variables is between 20 and 50. Where the number

of variables is less than 20, there is a tendency, not too serious, for this method to extract a conservative number of factors. When more than 50 variables are involved, too many factors are taken out.

(2) It will be recalled that component analysis has the drawback of containing 'hybrid' factors, particularly in the later factors to be extracted, because unique variance overlaps with common variance. Cattell has argued that some unique variance creeps into *all* factors and that the proportion in later factors to be extracted is so great as to

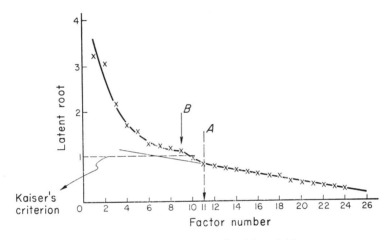

FIGURE 3.1. The scree test for 25 variables

swamp the common variance. We need to identify the optimum number of factors which can be taken out before the intrusion of non-common variance becomes serious. An intriguing method described by Cattell[9] is the *scree test*. For this, a graph is plotted of latent roots against the factor number (i.e. the order of extraction) and the shape of the resulting curve employed to judge the cut-off point.

Figure 3.1 gives a plot of the first 24 factors extracted for the writer in a recent study. Starting at the highest latent root, the plot is curved at first then develops into a linear relationship about point *A*. The point at which the curve straightens out is taken as the maximum number to be extracted. In this case the first 11 would qualify. Beyond 11 we have what Cattell described as 'factorial litter or scree' (scree, a geological term, is the debris which collects on the lower part of rocky slopes). Note the kink at *B* about two or three factors before the point *A* which seems fairly characteristic. Using the scree test gives us two more factors than Kaiser's criterion since the factor number with a latent root just greater than unity is the ninth (1·118) whilst the tenth latent root is 0·957. This means that the first nine would have been extracted using Kaiser's criterion.

CRITERIA FOR THE SIGNIFICANCE OF FACTOR LOADINGS

Ultimately we have to decide on which factor loadings are worth considering when it comes to interpreting the factors. Several methods have been suggested, of which three will be mentioned.

(1) The first suggestion is not really based on any mathematical propositions, except that it represents roughly 10 per cent of the variance. It is a rule of thumb which can be used in the first flush of excitement on making a preliminary inspection of the factor matrix. Some idea of the pattern of significant items can be gained by underlining the loadings greater than ±0·3 provided the sample is not too small (*N* = 50 at least). As we shall see, compared with other criteria, this is quite a rigorous level so we are not taking too much for granted.

(2) In deriving the factor loadings it became evident that they were, in effect, correlation coefficients. For the purposes of specifying an acceptable level of significance the loadings could be treated in a similar fashion to correlation coefficients. Appendix B has already been used for this purpose and loadings of at least ±0·11 and ±0·15 were recommended for the five per cent and one per cent levels when *N* = 300. Because of the uncertainty surrounding the assessment of error in factorial work, it would, perhaps, be safer to adopt the one per cent level as the criterion for significance. For large samples this criterion is much less demanding than the arbitrary figure of ±0·3. Researchers often take ±0·3 for the purposes of their main factor interpretations and apply the present or the following criterion where the variables possessing significant loadings are clearly meaningful.

(3) One distinct disadvantage in the last method is the absence of any adjustment for the number of variables, or the factor under consideration. Burt and Banks[10] have shown that as one progresses

from the first factor to higher factors the acceptable value for a loading to be judged significant should *increase* (it should get harder for coefficients to reach significance). Apart from anything else, the gradual intrusion of unique variance into later factors requires some adjustment in the level of significance. These authors devised a formula (the Burt–Banks formula) which has the merit of allowing not only for the sample size but also for the number of tests correlated and the number of factors extracted up to and including the one under examination. Appendix C gives the formula and some ready reckoned values for a selection of sample sizes.

It is of interest to note that the values of significant loadings for the first three factors are not far removed from those quoted in Appendix B. Moreover, as the factor number increases and the sample size decreases, the value required for significance rises sharply. For $N = 300$, at the one per cent level the first factor requires loadings of $\pm 0 \cdot 15$, factor II requires $\pm 0 \cdot 16$ and factor III requires $\pm 0 \cdot 17$ (see the formula in Appendix C).

The formula, in addition to providing some assessment of the standard error (S.E.) of a loading, can, according to Vernon,[11] serve to evaluate the common factors. First the standard error is calculated (Appendix C) then doubled. Only those factors possessing at least half the total number of variables with values in excess of $2 \times$ S.E. should be considered. The method works out to be exceedingly stringent, especially for small samples, and would, for instance, probably exclude factor III and certainly factor II in table 3.2 when using the one per cent level.*

INTERPRETING A FACTOR MATRIX

Using Kaiser's criterion for the number of factors to be extracted, reversing the signs where necessary to make the highest loadings positive and applying the arbitrary criterion of $\pm 0 \cdot 30$ to the factor loadings, table 3.2 becomes transformed to a more typical mode of presentation shown in table 3.3. This is one of the conventional

---

*Using Appendix C, $2 \times$ S.E. at the one per cent level for factors I, II and III are $\pm 0 \cdot 30$, $\pm 0 \cdot 32$ and $\pm 0 \cdot 34$ respectively. Only two loadings exceed $\pm 0 \cdot 32$ in factor II and only three loadings exceed $\pm 0 \cdot 34$ in factor III.

forms of presenting a factor matrix in research reports. Some reports omit the loadings which fall below significance in order to highlight the remaining ones. However, it is helpful and preferable to have all the loadings included if only to give the reader some idea of the signs and 'near misses'. The patterns of insignificant loadings can sometimes substantiate the interpretation of the significant loadings.

TABLE 3.3

Factor matrix for IQ and divergent tests
(Presented as in the research literature)

| Tests | Common factor loadings | | | Communality |
	I	II	III	(h^2)
1. AH5 verbal	32	82	01	7688
2. AH5 spatial	17	86	15	7870
3. Uses—F	76	−18	−07	6169
4. Uses—O	74	−05	−31	6354
5. Consequences—F	77	−02	−22	6402
6. Consequences—O	57	−05	−49	5603
7. Circles—F	66	−14	58	7824
8. Circles—O	51	−19	62	6788
Latent root	2·8551	1·4937	1·1210	5·4698

Decimal points omitted.
Significant loadings in bold (values greater than ±0·30).
F = Fluency; O = Originality.

In the first factor, all but the variable of spatial ability would qualify for consideration. A useful tactic is to consider the loadings in descending order of magnitude because those with the largest values are going to give us the 'flavour' of the factor. The backbone of this first factor consists of the divergent thinking tests with verbal IQ at the tail end. If we had set up a hypothesis that there should be a general intellective component embracing all the tests in the example, it would have been rejected. The well-established measures of intelligence, such as the AH5, have not figured sufficiently prominently in the first factor for us to conclude that it is a general factor of intellectual capacity. Logically, factor content is a function of the variables selected for examination and the rather small number of standardized intelligence tests in the present case may have affected the factor structure.

We could make all manner of speculations about the label to be ascribed to factors—an exciting though perilous business with inadequate test material as in the present example. Factor I, for instance, might be called a 'verbal ideational fluency' factor in drawing together the divergent thinking tests and the verbal AH5. In other words, we are guessing that speed and variety in the generation of ideas is the important common element which correlates these tests. But rather than struggle to find labels for these principal components, it might be as well to wait until we have had a chance to apply other kinds of factorial solutions to be described in the next chapter.

Returning to the correlation matrix in table 3.1 for a moment will reveal some close similarities between the general picture of significant (and sizeable) correlations and the first factor. Note that all but the spatial AH5 values in column 2 are positive and of modest size and this is reflected in the exclusion of the spatial variable in factor I. Superficially it seems that the first factor represents this general positive interrelationship between the variables.

The second factor loads on the IQ tests and probably accounts for the high correlation between these variables. The fact that we have been able to relate the correlations to the factor matrix so well is partly fortuitous in that the patterns of the former are conspicuous. Correlation matrices are not always so revealing. Factor III is particularly interesting because it contains significant loadings which are both positive (tests 7 and 8) and negative (tests 4 and 6). This arrangement is known as a *bipolar factor* because the factor embodies contrasting groups of variables. In geometrical terms, some of the test vectors have been resolved in one direction and others in opposite quadrants to give rise to positive and negative values. Provided we are consistent, it does not matter which direction is designated as positive. Thus in factor II we could have made tests 1 and 2 negative provided we had reversed the signs of tests 3 to 8 to make them positive. This has been done in factor III to make the highest loadings positive (compare tables 3.2 and 3.3).

A bipolar factor can be labelled in one of two ways. Either it is described in terms of the significant variables at the poles of the factor or in terms of a single nomenclature which appropriately describes the continuum from one pole to the other. In factor III, 'Circles' appears at one pole and the originality scores of the remaining divergent tests at the other. The bipolar factor might be called a non-verbal/verbal originality dimension. In this case a label describing the continuum would not seem to be appropriate. A more obvious example illustrating both descriptions is afforded by the extremes of, say, brightness and darkness (a bipolar description) or light intensity (a continuum).

Before making any statements about the causal relationships in the three factors, additional supporting evidence would certainly be necessary. Apart from anything else, more tests would be highly desirable. There is also a certain amount of confusion in factors I and III with no particularly sensible or familiar patterns emerging. Furthermore, the second factor might easily have been discarded had Vernon's method been applied to establish the worthwhile factors. As this factor contributes approximately 19 per cent of the total variance and adequately satisfies Kaiser's criterion, there is sufficient justification for retaining it in subsequent analyses to be described in the next chapter. In it we shall discuss the attempts made by behavioural researchers to restructure factor matrices, such as the one shown in table 3.3, in the hope of deriving greater meaning and usefulness from the factor pattern.

REFERENCES

1. For a recent discussion of the issues see J. Freeman, H. J. Butcher and T. Christie, *Creativity: a Selective Review of Research*, Research into Higher Education Monograph, London, 1968.
2. This test is a measure of high-grade intelligence devised by A. Heim and published by National Foundation for Educational Research in England and Wales, London.
3. Many divergent thinking tests have been constructed by Guilford in the United States. A number of these have been adapted and published in E. P. Torrance, *Guiding Creative Talent*, Prentice-Hall, New Jersey, 1962. 'Uses of Things' appears in J. W. Getzels and P. W. Jackson, *Creativity and Intelligence*, Wiley, London, 1962.

4. Details of modifications for British samples and scoring techniques can be found for these tests in D. Child, *A Study of some aspects of divergent thinking and their relationship to arts and science preferences in schoolchildren*', M.Ed. Thesis, University of Leeds, unpublished, 1968.

5. Child (1968).

6. R. B. Cattell, *Factor Analysis*, Harper, New York, 1952, discusses the finer points of correlation techniques and their applicability in factor methods.

7. H. J. Hallworth and A. Brebner, *A System of Computer Programs for Use in Psychology and Education*, British Psychological Society, London, 1965.

8. For example, Cattell (1952), Harman (1967).

9. R. B. Cattell, *Handbook of Multivariate Experimental Psychology*, Rand McNally, Chicago, 1966, pp. 174–243. An application appears in 'The theory of fluid and crystallized intelligence', *Br. J. educ. Psychol.*, 37, 209–224 (1967).

10. C. Burt, 'Tests of significance in factor analysis', *Br. J. Psychol.*, 5, 109–133 (1952). In addition to the Burt–Banks formula, this article gives a thorough ventilation of the theme of factor significance.

11. P. E. Vernon, *The Structure of Human Abilities*, Methuen, London, 1950, p. 130.

CHAPTER FOUR

THE ROTATION OF FACTORS

DIRECT AND DERIVED SOLUTIONS

The methods of analysis described so far are sometimes referred to as *direct* methods because the factor matrix obtained arises directly from the correlation matrix by the application of mathematical models. It was stated elsewhere that these orthogonal reference axes were by no means binding, but many researchers use the results of centroid or principal components programmes for the final interpretation of the correlation between variables. This was briefly attempted in the last chapter using the principal components in table 3.3. But do these tidy orthogonal factors, whilst fulfilling desirable mathematical requirements, invariably produce information which offers the most adequate interpretation of the variables under examination? Are the reference axes from direct solutions always in a position to give the most illuminating evaluation of the variables? Mathematically equivalent designs can, and do, give rise to a variety of alternative and equally acceptable solutions. Is there some way of producing a compromise between them which might give invariant solutions?

Most factor analysts are now agreed that some direct solutions are not sufficient. In most cases, adjustment to the frames of reference improves the interpretation by reducing some of the ambiguities which often accompany the preliminary analysis. Sometimes modifications in the position of the frame of reference provide solutions which square more convincingly with independent and non-factorial evidence. This point will be raised again when considering the work of Eysenck and his 'criterion analysis' in chapter 5. The process of manipulating the reference axes is known as *rotation*, and this will be the subject of the present chapter. The results of rotation methods are sometimes referred to as *derived solutions* because they are obtained as a second stage from the results of direct solutions.

Before dealing with rotation, however, an observation about direct methods needs to be mentioned. Referring back to figure 2.7 and table 2.5 in chapter 2 will reveal that the second and subsequent factors are bipolar. These bipolar factors occur frequently in all but the first factor extracted because of the very nature of the analysis. With the second factor vector passing between the test vectors, the second factor vector, at right-angles to the first, will inevitably give rise to negative loadings. In many cases this will give both positive and negative significant loadings. The same argument applies to most of the subsequent factors extracted. Some workers see little virtue in taking these direct solutions alone for interpretation because, as they say, incidental mathematical features inherent in the procedures are being used as the grounds for psychological theories. Guilford,[1] for one, has always been highly critical of *g* theory in the study of human ability because, he claims, the presence of a major share of common variance in the first factor of a direct method is a function of factorial design and not necessarily a structural feature of human ability. Whether this view is true or not, it is advisable to guard against reading too much into preliminary analyses without first exploring derived results, if possible, alongside non-factorial evidence.

THE NAMES OF SOME DIRECT METHODS

A number of equivalent and acceptable direct solutions have evolved over the years. The reader of research literature might come across them referred to by name. Therefore, a list of some of the commonest methods should prove helpful for identifying the stage of analysis involved. Principal Components or Axes (Pearson and Hotelling), principal factor (Thomson), centroid or simple summation (Thurstone and Burt) and bi-factor designs (Holzinger) need to be mentioned. In all these, some estimate of the communality can be adopted as an alternative to unity. However, another group of methods has arisen in which an estimate of the number of common factors is the basic requirement for initiating factor extraction. The communalities in these methods emerge incidentally because the number of factors taken out is fixed. Two such methods which have only gained ground in recent years, primarily because they require long-winded calculations more suited to computer methods, are the method of 'maximum-likelihood' enunciated by Lawley[2] and the 'Minres' solution devised by Harman.[3] As the latter points out in his book *Modern Factor Analysis*, the solutions are 'dependent upon an estimate of the number of common factors; the communalities consistent with this hypothesis, are obtained as by-products of the method.' These methods have also the added advantage that the adequacy of the number of factors speculated at the outset can be tested with statistical precision after the analysis.

ROTATION (OR MULTIPLE-FACTOR SOLUTIONS)

The term 'rotation' applied to the reference axes means exactly what it says, namely, the axes are turned about the origin until some alternative position has been reached. The simplest case arises when the axes are maintained at 90° thus giving an *orthogonal rotation*. Further, it is quite possible, and more popular, to rotate the axes through different angles to arrive at an *oblique rotation*. We will demonstrate the orthogonal procedure using a graphical approach, an approach which is still used by a few researchers in the final stages of a rotation. Nevertheless, hand rotation is a laborious technique and the growth of computer facilities has encouraged most researchers to rely on 'analytical' rotation, that is, obtaining computerized solutions using mathematical approximations.

First, a simple example of the effects of rotation will be given. Suppose that the loadings of a test in the first and second factors (F_1 and F_2) of a direct analysis are $+0.40$ and $+0.80$ respectively. Retracing our steps from a knowledge of loadings to the position of the vectors, we obtain figure 4.1(a). In this case, our orthogonal coordinates are F_1 and F_2, whilst OA is the test vector. A is located using the loading values on the appropriate axes.

Imagine a pin stuck through the origin O in figure 4.1(a) preventing its movement and the axes OF_1, OF_2 free to rotate through any required angle to a new position OF_1^1, OF_2^1 *with the point A stationary*. For simplicity, let the angle of rotation be 30° thus giving the situation shown in figure 4.1(b). What are the new values of the point A on OF_1^1, and OF_2^1 at the perpendicular projections from A

onto these axes? For a 30° rotation these values, represented by B and C, are 0·75 and 0·50. The values of OB and OC can also be

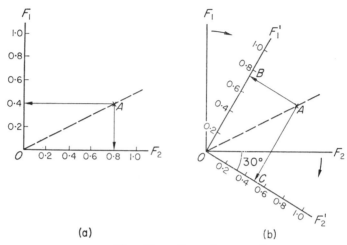

(a)   (b)

FIGURE 4.1. The effect of rotating coordinates

determined using a standard formula. For the general case, if θ is the angle of rotation, OB is given by

$$OB = F_1 \cos \theta + F_2 \sin \theta \qquad (1)$$

where F_1 and F_2 are the initial loadings. In the present example, $\theta = 30°$, $F_1 = 0·40$ and $F_2 = 0·80$

$$\therefore OB = 0·4 \cos 30 + 0·8 \sin 30$$

hence

$$OB = 0·4 \times 0·8660 + 0·8 \times 0·5$$

which gives

$$OB = 0·3464 + 0·4000 = 0·7464$$

Similarly,

$$OC = F_2 \cos \theta - F_1 \sin \theta \quad \text{(note the sign change)} \qquad (2)$$

which gives

$$OC = 0·80 \times 0·8660 - 0·40 \times 0·5$$

that is

$$OC = 0·6928 - 0·2000 = 0·4928$$

These values for OB and OC compare with the graphical solution when we allow for rounding errors.

An important question must be, has this modification affected in any way the common variance of the test? Before rotation, the variance of F_1 and F_2 would be the sum of their squares, that is

$$(0.40)^2 + (0.80)^2 = 0.16 + 0.64 = 0.80$$

For the rotated position, the variance is

$$(0.7464)^2 + (0.4928)^2 = 0.56 + 0.24 = 0.80$$

The common variance is identical and would remain so irrespective of the position of the coordinates, provided the rotation occurs about the origin O, and the test vector OA remains stationary. This follows from the geometry of the arrangement, for the length OA, which is that portion of the communality accounted for in factors 1 and 2, is fixed and equal in the existing case to 0.80. However, this variance has been redistributed by the act of rotation such that the loading in the first factor has changed from 0.40 to 0.7464, and in the second factor from 0.80 to 0.4928. We shall apply the above process to an actual rotation problem later, but first we have to discuss the criteria which help in judging when to halt the rotation for the most efficacious position of the axes.

THURSTONE'S THEORY OF SIMPLE STRUCTURE

The earliest attempts at rotation came in the 1930's when Thurstone expounded his theory of 'simple structure'. His primary objective was to organize the factor axes (and hence the loadings) so that their meaning would make better sense in psychological terms. In his particular case, he was concerned with the structure of mental abilities.[4] For him, direct solutions, which satisfied the principle of parsimony in reducing a large number of related variables to a small number of independent factors, were not adequate. In addition to parsimony, solutions should be invariant, unique and in accord with non-factorial research findings. He believed that factor analysis was most appropriately used as the first stage in mapping out new domains and not as an end in itself. The findings of factorists was seen as the starting point for non-factorial experiments.

By invariance, Thurstone was referring to the constancy of factor content from one analysis to the next. Where a factor for numerical skill appears, for example, this should be regularly in evidence whenever the same or a similar test battery is used with samples of the same population. Uniqueness occurs when the resulting model is singularly appropriate for the description of the underlying causes of a factor. If the same domain is investigated, identical configuration of the factors should be present. The problem facing Thurstone was to find a mathematical design which enabled him to discover 'the unique solution'. The problem still exists!

In an attempt to fulfil the requirements of unique and invariant factors, Thurstone established several criteria to assist in the decision as to when rotation should cease. The criteria were intuitive and are not rigidly adhered to nowadays, although they still lurk in the background of most subsequent formulations. They are based on the principle that the simplest explanation involving only a few variables is the best. The fullest statement appears in his book on *Multiple Factor Analysis* in which he proposes five conditions for the fulfilment of simple structure for an orthogonal or oblique analysis. If we take a factor matrix from a direct evaluation, the derived matrix after rotation should meet the following requirements.

(1) Each row of the derived matrix, that is the loadings associated with one variable, should contain at least one zero loading. A zero loading would include numerical values which were not statistically significant.

(2) If there are n common factors being used in the rotation (selected using one of the criteria for deciding on the number of worthwhile factors as indicated in the last chapter) there should be at least n zero loadings in *each* factor.

(3) For every pair of factors there should be several variables with zero loadings in one factor but having at the same time significant loadings in the other.

(4) For every pair of factors a large proportion of the loadings should have zero values in both factors where there are four or more factors.

(5) For every pair of factors there should only be a small proportion of loadings with significant values in both factors.

These criteria have the effect of maximizing the number of loadings having negligible values whilst leaving a few with large loadings. As we shall see, this makes the job of interpreting a factor very much easier than would a collection of moderately sized loadings.

FACTOR ROTATION (ORTHOGONAL)

Let us look at a rotation using familiar results from a previous chapter. The least complicated method is an orthogonal rotation and we shall apply this to the Principal Components analysis portrayed in table 2.5. The table is reproduced in figure 4.2(a) to two decimal places only. The second diagram shows the arrangement when the loadings of factor I are plotted against factor II, the factors acting as reference vectors. Notice that the points in figure 4.2(b) are identical to the configuration of test vectors displayed in figure 2.7.

| | Factor loadings | |
	I	II
Test 1	57	−82
Test 2	71	−71
Test 3	97	26
Test 4	82	57
Test 5	71	71

(a)

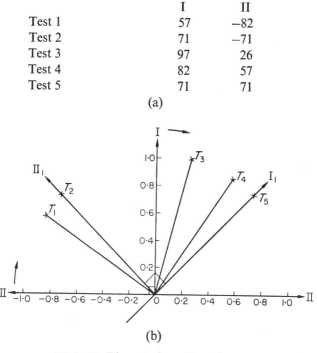

(b)

FIGURE 4.2. The rotation of two factors

The next stage is to rotate the reference axes I and II, clockwise in the present example, keeping them at 90° to each other until by inspection as many points as possible come close to *the plane at right-angles to the reference axis.* As noted above, Thurstone's simple structure requires as many as possible of the variables to give negligible loadings, which means that most of the points should fall close to an axis (in the plane of the paper in two-dimensional arrays) or a plane (in three- or more dimensional arrays) at right-angles to the reference vector. These are known as *hyperplanes.* Clearly, if points come close to this plane, the projection onto the reference axis will be close to zero. In figure 4.2(b), a two-dimensional arrangement, rotation of I to I_1 has brought T_1 and T_2 close to the hyperplane of I_1 which so happens to be the new position of II_1 in an orthogonal rotation. By alternately determining the projections of T_1 to T_5 onto the new reference axes and adjusting their position, we are aiming to approximate to a position which most satisfies simple structure. In orthogonal arrangements, the projections onto the second factor vector must be considered simultaneously because the new position of the reference vector I is also the hyperplane of reference vector II. In the existing example the rotation was stopped when the hyperplane of factor I passed through the point T_2, and since T_2 and T_5 are 90° to each other, the factor II hyperplane will pass through T_5. This position was taken as a first approximation and as the most mathematically convenient. Some position between T_1 and T_2 might have been more satisfactory for simple structure. Table 4.1(a) illustrates the derived loadings for the new reference axes and the results of a Varimax* rotation for the same variables in 4.1(b).

The common variance for each test has remained constant despite alterations in the loadings. The high and low loadings are somewhat more conspicuous than in the direct matrix reproduced in figure 4.1(a). The Varimax solution is a better solution than the author's primarily because it has located a position for the new reference axes which tends to eliminate the ambiguity posed by the highish loading of test 3 in factor II_1. This has been achieved by

---

*The Varimax method was devised by Kaiser and enjoys extensive usage in both America and this country.

TABLE 4.1

Rotated factor loadings

	(a) Graphical solution			(b) Varimax solution		
	I_1	II_1	h^2	I_1	II_1	h^2
Test 1	−0·1736	0·9848	1·0000	−0·033	0·996	0·993
Test 2	0·0000	1·0000	1·0000	0·141	0·987	0·976
Test 3	0·8660	0·5000	1·0000	0·914	0·368	0·971
Test 4	0·9848	0·1736	1·0000	0·999	0·035	0·999
Test 5	1·0000	0·0000	1·0000	0·989	−0·138	0·997

taking the rotated axes to positions somewhere between T_1 and T_2 for one and T_4 and T_5 for the other orthogonal coordinate. More experienced hands and eyes, can make these graphical hand rotations a most effective technique.[5] Observe that the communalities in the Varimax solution are less than one principally because an estimate equal to the largest correlation for each variable was inserted in the relevant leading diagonal, instead of unity, before the rotation was commenced. The method has been mentioned in the last chapter. Notice how neatly the rotated factor vectors run between T_1 T_2 and T_3 T_4 T_5 for the Varimax solution. Clearly, the factor vectors have tended to locate the clusters of test vectors. An oblique solution would have ignored the 90° angle between the factor vectors and would probably come close to giving resultants falling exactly between T_1 T_2 and T_3 T_4 T_5. Note also that all but one of Thurstone's criteria for simple structure have been satisfied. The loadings of test 3 are significant in both rotated factors and there should ideally have been one zero loading.

With three or more factors to deal with, the corresponding graphical or analytical solution would be performed in a number of stages depending on the number of factors involved. As each axis must be rotated in turn with all other axes, the task of rotating even a moderate number of factors can be a marathon project. For three factors, there are three operations and for six factors, fifteen are required. Let the original orthogonal axes for three factors be X_1 Y_1 and Z_1. The first step would be to rotate X_1 Y_1 to new positions X_2 Y_2. The other factor, Z_1, is now rotated with one of the factors in a new position, say X_2, to give $Z_2 \, X_3$. Finally, the axis Y_2 now needs to be rotated with Z_2 to give every possible combination. Let these new axes be $Y_3 \, Z_3$. Manipulating even three axes is not an easy task and Cattell's (1952) outline of the mechanical procedures leaves one in no doubt about the difficulties involved.

A NOTE ON SECOND- AND HIGHER-ORDER FACTORS

The oblique case is rather more complex than the orthogonal. No entirely satisfactory analytical rotations have been devised for oblique solutions and they are still the subject of considerable experimentation and controversy. Cattell (1952) is a keen protagonist of the method and he has made a significant contribution in this direction. His explorations have, in fact, gone beyond the first oblique analysis to 'second- and higher-order' factors.* As the factors of the initial oblique rotation (first-order analysis) are correlated, it is quite conceivable that these correlations between the first-order (or primary) factors can be treated as a correlation matrix for the purposes of a second factor extraction and oblique rotation. The resulting factors of factors are of the second order (actually an 'inverse matrix' is used rather than the direct correlations, but the details need not concern us here). Exactly the same process can be adopted in deriving higher-order factors. Obviously, each order reduces the number of factors of the preceding order until there remain only one or two factors—often at the fourth- or fifth-order of extraction.

As the term 'oblique' implies, there would certainly exist a correlation between the factors extracted. Oblique rotation with behavioural variables is an admission that most, if not all, human characteristics are correlated to some extent, and the underlying major factors must be similarly correlated. The controversial issue concerns the difficulty in deciding the extent of this correlation.

One growing point of factor analysis over the next few years will certainly be in the direction of establishing analytical solutions for both orthogonal and oblique cases, particularly for the simultaneous rotation of factors. As we have seen in the last paragraph, rotations

---

*For an illustration see the section relating to Cattell's work on fluid and crystallized intelligence in chapter 5.

possessing more than two factors are carried out two at a time and this has some slight effect on the results when compared with simultaneous rotation.

THE NAMES OF SOME DERIVED METHODS

The expansion of analytical solutions, almost entirely the fruits of computerization, started in the early 1950's. Harman[6] says:

> From the very beginning of the application of simple-structure principles, it was recognized that the procedure was more of an art than a science. In an endeavour to put the rotations on a more objective basis, the first improvements were directed towards eliminating graphical procedures. (p. 294)

It was Carroll[7] who provided the first 'truly analytical rotation for determining psychologically interpretable factors' (Harman, p. 295) by modifying some of the simple structure criteria laid down by Thurstone. The method has become known as the quartimax version and can be used to solve orthogonal problems. To assist the reader in his reading of research literature, a short list of derived methods and associated originators is included. Orthogonal solutions are possible using Varimax rotation (Kaiser). Oblique solutions bear such names as Quartimin (Carroll), Oblimin (Carroll), Covarimin (Kaiser), Biquartimin, Binormamin, Oblimax, Promax, Procrustes and Maxplane (Cattell).

Harris[8] has recently advanced an interesting proposal that several rather than one factor programme should be employed with a given collection of data; this, he points out, should test the 'robustness' of the factors. By robustness he means the regularity with which particular factors reappear for homogeneous cases irrespective of the analytical techniques adopted. By selecting a few methods of analysis which differ in principle, comparing the results of the rotated factors from these methods and retaining only those factors which persist whatever the method, one might arrive at a 'maximum common denominator' for substantial dimensions within an experimental domain.

ROTATION OF THE IQ AND THE DIVERGENT TEST MATRIX

To illustrate the application and interpretation of a rotated matrix, we shall return to the IQ and divergent test correlation matrix in table 3.1. The Principal Components analysis proved valuable in deciding how many factors to retain for rotation. With unities in the diagonals, only three of the factors extracted had latent roots greater than one. The rotation programme used in the study was the Varimax procedure[9] which gives an orthogonal solution.

TABLE 4.2

Varimax analysis of IQ and divergent tests (orthogonal)

| Tests | Rotated factor loadings | | | Communality (h^2) |
	I	II	III	
1. AH5 verbal	17	**86**	01	7688
2. AH5 spatial	−04	**89**	04	7871
3. Uses—F	**68**	−02	**39**	6169
4. Uses—O	**78**	07	16	6354
5. Consequences—F	**75**	12	24	6402
6. Consequences—O	**74**	01	−08	5604
7. Circles—F	23	07	**85**	7824
8. Circles—O	08	−01	**82**	6788
Percentage variance	28·53	19·37	20·47	68·37

Decimal points omitted.
Significant loadings in bold.
F = Fluency; O = Originality.

To be on the cautious side, let us use the arbitrary criterion of ±0·30 for the significance of factor loadings. These are in bold type. Note that the total percentage variance is precisely the same as in the Principal Components analysis of table 3.2 because changing the position of the reference vectors does not alter the total variance. This would only occur if we altered the communality prior to rotation, a method sometimes adopted when component analysis is used to establish the number of factors to be extracted with unities in the leading diagonals followed by an analysis using an estimate of the communality. Also observe that the communalities for each variable are unaltered. What has changed is the distribution of variance between the three factors. For example, comparing the AH5 verbal loadings in table 3.2 with table 4.2 reveals a transfer of some variance from the first to the second factor (by moving the reference vector). As we might have expected from the discussion of rotation, the loadings of moderate size have, in the main, become larger, or smaller to the point of insignificance. This has happened with the loading of AH5 verbal on the first factor, having changed from 0·32 to 0·17 (insignificant) after orthogonal rotation.

These factors now look quite promising. Several marginally significant loadings in the Principal Components analysis have been

TABLE 4.3
Promax analysis of IQ and divergent tests (oblique)

Tests	Rotated factor loadings			Communality (h^2)
	I	II	III	
1. AH5 verbal	12	**86**	−04	75
2. AH5 spatial	−11	**89**	03	81
3. Uses—F	**64**	−07	27	49
4. Uses—O	**79**	02	00	63
5. Consequences—F	**75**	07	09	57
6. Consequences—O	**82**	−03	−25	73
7. Circles—F	05	04	**86**	74
8. Circles—O	−10	−04	**86**	75
Percentage variance	28·79	19·30	20·28	68·37

Decimal points omitted.
Significant loadings in bold.
F = Fluency; O = Originality.

reduced to insignificant levels and a few sizeable loadings remain. All Thurstone's postulates for simple structure have been satisfied. In the first factor the verbal divergent thinking tests (Uses and Consequences) have been isolated. The second factor loads specifically on the AH5 variables. Factor III loads substantially on the Circles test which requires non-verbal divergent thinking skills, and the fluency element of the Uses test. Rotation has certainly helped to clarify the pattern of tests as compared with the matrix of table 3.2. At least we have factors which contain certain recognizable combinations of variables making good sense in terms of current research in this field. A glance at table 3.1 showing correlations confirms the major groupings of variables within each factor.

An oblique rotation was also carried out using these data by a method known as a Promax solution.[10] Table 4.3 shows the outcome. The total variance is the same as before, but in abandoning the orthogonal structure and replacing it with vectors which correlate, we find that neither communalities nor percentage variance for each factor can be identified with the values obtained in the Principal Components and Varimax matrices. The factor pattern is not unlike the orthogonal structure in table 4.2 except that the loadings have been pushed even higher or lower. Thus, the significant variable 3 in the third factor of table 4.2 has now become insignificant if we are using ±0.30 as the significance level. The resulting picture separates quite sharply the verbal ideational fluency tests in factor I, the conventional

TABLE 4.4

Correlations between Promax factors

	1	2	3
1.	1·0000	0·1346	−0·3991
2.	0·1346	1·0000	−0·0727
3.	−0·3991	−0·0727	1·0000

tests of verbal and spatial ability in factor II and the non-verbal ideational fluency test in factor III. But it is important to remember when interpreting oblique resolutions that the factors may well be correlated, and table 4.4 illustrates this for the present example.

If the coefficients were converted to cosines and a vector model drawn, some idea of the interrelationship between the factors could be demonstrated. The correlation of −0·40 between factors 1 and 3 is not really surprising when we look back at table 3.3 to discover all the divergent thinking tests in one factor. As will be seen in the next chapter when exploring Cattell's work on intelligence, the overlapping factors converge in higher-order analyses to make them more all-inclusive.

Finally, a word might be said about the presentation of findings in research journals. Most readers of this work may never experience the exciting task of initiating, interpreting and reporting the results of a factor analysis. But for those who do, this is a plea for their reports to provide sufficient information for others to make an intelligent assessment of the findings. A number of reports leave far too much to the imagination. Apart from showing all the relevant matrices with latent roots and communalities where necessary, it would be of great assistance if journals established basic requirements for all factor analysis articles. There are obviously many useful elaborations which factorists may wish to include, but a bare minimum should consist of (a) a statement of the direct method used, (b) details about the entries in the leading diagonals, (c) the criterion for deciding on the number of factors to be extracted, (d) the criterion for choosing the significant loadings in each factor and (e) the rotation method adopted, if relevant. These would form a very acceptable baseline of information.

REFERENCES

1. J. P. Guilford, in *Handbook of Measurement and Assessment in the Behavioral Sciences* (Ed. D. K. Whitla), Addison-Wesley, Reading, Mass., 1968, has some harsh points to make about the *g* theory of human ability.
2. D. N. Lawley, 'The estimation of factor loadings by the method of maximum-likelihood', *Proc. R. Soc. Edinb.*, **60**, 64–82 (1940).
3. H. H. Harman and W. H. Jones, 'Factor analysis by minimizing residuals (Minres)', *Psychometrika*, **31**, 351–368 (1966).
4. Thurstone's ideas are developed in 'Multiple factor analysis', *Psychol. Rev.*, **38**, 406–427 (1931). Brief mention is also made of his 'Primary Mental Abilities' in chapter 5. See also *Multiple Factor Analysis*, University of Chicago Press, Chicago, 1947.
5. For a lucid description of both orthogonal and oblique rotation, the reader is strongly urged to read R. B. Cattell, *Factor Analysis*, Harper, New York, 1952. He devotes a good deal of space to the methods and problems of the procedure.
6. H. H. Harman, *Modern Factor Analysis*, University of Chicago Press, Chicago, 1967.
7. For a useful discussion of recent trends, see a paper by F. W. Warburton, 'Analytic methods of factor rotation', *Br. J. math. statist. Psychol.*, **16**, 165–174 (1963). Carroll's own work appears in 'An analytic solution for approximating simple structure in factor analysis', *Psychometrika*, **18**, 23–38 (1953). See also H. H. Harman (1967).

8. C. W. Harris, 'On factors and factor scores', *Psychometrika*, **32**, 4, 363–379 (1967).

9. H. F. Kaiser, 'Computer program for Varimax rotation in factor analysis,' *Educ. Psychol. Meas.*, **19**, 413–420 (1959).

10. A. E. Hendrickson and P. O. White, 'Promax: a quick method for rotation to oblique simple structure'. *Brit. J. math. statist. Psychol.*, **17**, 65–70 (1964).

CHAPTER FIVE

SOME APPLICATIONS

The variety and complexity of factorial studies over the past thirty years has been tremendous. Even a summary of the scope of research would be an impossible task. In this final chapter we shall have to be satisfied with a meagre selection of illustrations from psychology, sociology and medicine which should serve to demonstrate a few interesting directions which research has taken. The choice of some aspects of human ability and inferential studies of personality is obvious, for it is in these areas where the method has had a substantial impact on the theories generated. Moreover, the reader will probably be familiar with the background in these subjects. The origin of Osgood's Semantic Differential, a comparative newcomer, has been included because it has recently enjoyed wide usage particularly as a measure of attitudes. An example is also taken from experimental psychology using factor analysis explicitly for descriptive purposes. An interesting theory regarding the way in which people identify themselves in relation to their work provides an example of factorial design in sociology. And, finally, a small selection of studies is taken from medicine and public health where there appears to be a growing application of factor methods.

HUMAN ABILITY

Of all human activities, intelligent behaviour, not surprisingly, has produced an abundance (possibly an excess) of research and controversy.[1] Factor analysis has flourished as a by-product of the psychologists' quest to disentangle man's intellectual make-up. One can trace a great deal of the development of factorial prescriptions simply by referring to studies on the nature of intelligence. In this section, only the outline of important models derived from factorial designs will receive attention.

Spearman's Two-Factor Theory[2] is regarded as the first formulation about the structure of human ability relying for its justification on factor analysis. His view was rooted in the assumption that 'general' ability (g) accounted in substantial measure for differences in human performance and was held to be innate. Ability in the

Test	\multicolumn{6}{c}{Factor}
	I	II	III	IV	V	VI
A	√	√				
B	√			√		
C	√				√	
D	√		√			
E	√					√
	general factor			specific factors (one for each test)		

√ represents a significant loading, and blank spaces indicate zero or negligible loadings.

FIGURE 5.1. Two-factor profile

'eduction of relations and correlates' was the quality which differentiated those with high and low g. Further, each test required a specific ability (s) which accounted for the unevenness in an individual's performance from test to test and was influenced by education. A boy, therefore, with high general ability would be expected to perform well in all school subjects because of his high mental efficiency. At the same time, differences in his performance between school subjects would arise from his special talents.

To support this theory from factor solutions would require one all-embracing common factor with high loadings in all items which test ability in the 'eduction of relations and correlates', and a series of factors, as many as there are tests, with a single significant loading in each—one for every test. Figure 5.1 is a popular way of summarizing the appearance of the significant loadings and factors.

This extreme example of the principle of parsimony in interpreting the cognitive domain appealed to many psychologists of the time and it enjoyed widespread support. However, to obtain the format in figure 5.1 requires certain assumptions. Spearman's selection of tests was such as to exclude group factors, that is small groups of tests having common characteristics. Obviously, by selecting

Test	Factor									
	I	II	III	IV	V	VI	VII	VIII	IX	X
A	√	√			√					
B	√	√	√			√				
C	√			√					√	
D	√		√					√		
E	√	√		√						√
F	√			√			√			

general factor ←—— group factors ——→

common factors ←————————→ ←———— specific factors ————————→

FIGURE 5.2. Hierarchical group-factor profile with overlapping group factors

a single representative test from diverse groups, no group factors could possibly emerge. Adding to this the theoretical suppositions[3] adopted, he was able to produce a pattern similar to figure 5.1. Hence, by judicious test selection and analysis peculiar to his researches, Spearman had obscured the other form of common factor, namely the group factor alluded to in chapter 3.

Burt, as far off as 1917,[4] was among the first to criticize Spearman's approach. Larger samples of subjects* and test materials

---

*Spearman, it appears, carried out some of his earliest investigations with small samples of $n = 24$ and $n = 22$, thereby increasing the need for corrections for unreliability.

alongside his simple summation method (see Appendix A) for simplifying the tedious calculations involved in component analysis enabled group factors to be explored. Psychologists were soon occupied discovering and isolating groups such as verbal, numerical and practical factors. Figure 5.2 gives an idea of the factor profile obtained.

It is possible, as Holzinger[5] has shown, to derive group factors which do not overlap (the bi-factor approach). In figure 5.2 test B has significant loadings in both group factors II and III. A solution is possible where no single test appears more than once in the group factors.

All the methods so far described have been *direct* solutions which have led factorists in Great Britain, headed by Burt, to postulate the *hierarchical theory* of the structure of human abilities.[6] The theory is too well known to necessitate a detailed elaboration here. Briefly, if an extensive battery of tests covering most of the human abilities (of which we are aware) is given to a large (thousands rather than hundreds), representative sample of the population, the factors from a direct analysis (centroid, simple summation, Principal Components) appear as a hierarchy. The first factor accounts, as usual, for a major portion of the common variance and involves all the tests of human ability; this is the g factor. Following the extraction of the g factor, two major group factors of verbal, numerical and educational skills on the one hand (v : ed factor) and practical, mechanical and spatial on the other (k : m factor) are discernible. With a sufficiently wide variety of tests and sufficient (at least three) of each variety, the major groups subdivide to give minor group factors. As usual, specific factors may then emerge to account for some of the specific variance. The overall picture, crudely, is a tree-like structure with g as the trunk, main branches as major factors, subsidiary branches as minor factors and twigs as specific factors.

Another milestone in the development of cognitive ability analysis came in the 1930's when Thurstone in America expounded his multiple-factor theory as an outcome of orthogonal (and later oblique) rotation referred to in the last chapter. The common variance, as we have seen, is spread out amongst the common factors instead of being concentrated in the first factor. The common variance of a test in these circumstances tends to be confined to a few common group factors. Several clear group factors were isolated and Thurstone referred to these as the *Primary Mental Abilities*.* He recognized the existence of a general factor given particular modes of analysis, but he found little use for the concept.

Figures 5.2 and 5.3 highlight the major distinction between the British and American psychologists' approach to the nature and structure of mental ability. Starting with precisely the same data, but different assumptions about the structure of ability, one could end up with at least two solutions by choosing either a direct or a

Test	Factor									
	I	II	III	IV	V	VI	VII	VIII	IX	X
A	√			√	√					
B	√		√							√
C	√	√					√			
D		√	√						√	
E		√		√				√		
F				√		√				

$\longleftarrow$ group factors $\longrightarrow$ $\longleftarrow$ specific factors $\longrightarrow$

FIGURE 5.3. Multiple-factor profile

derived technique. It is important to remember from our previous arguments that these different approaches are related; the multiple group factors of Thurstone could be 'derived' from the general and group factors of the hierarchical approach. The debate between the schools of thought in Britain and America hinges on disagreements about the appropriateness of the mathematical models employed, the psychological meaning of the results obtained and the use to which these results can be put.

---

*Examples of Primary Mental Abilities are verbal comprehension, number skills, word fluency, perceptual flexibility and speed, inductive reasoning, rote memory and deductive reasoning.

Recently, Cattell (1963 and 1967, for example)[7] has advanced a theory of intelligence in which two general factors are postulated, namely, *fluid* and *crystallized* general ability.

According to the theory of fluid and crystallized general ability, there is not *one* general ability factor, as originally propounded by Spearman (1904) and supported by Thurstone (1938), but *two*. It states that these two broad factors are distinguishable by one, called crystallized intelligence, g_c, loading most heavily the culturally acquired judgmental skills, while the other, called fluid ability, g_f, is found loading insightful performance in which individual differences in learning experience play little part. (1967)

| Test | First-order factors | | | | Second-order factors | | |
	I	II	III	IV	First-order factors	I	II
A			√				
B	√	√					
C	√	√		√			
D	√			√	I	√	
E		√			II		√
F	√				III		√
G				√	IV	√	√
H	√			√			
I			√			g_f	g_c
J			√				
	g_f	←———group———→ factors (primary abilities)					

(a)   (b)

FIGURE 5.4. Cattell's fluid-crystallized general abilities

Earlier work, according to Cattell, did not expose the double structure because factor techniques were cruder, particularly in the estimation of communalities and the number of significant factors worth extracting. Perhaps the most important defect was the sparseness of 'hyperplane stuff', that is the variety and quantity of test material involved in the factor analysis. With the inclusion of material extending across the cognitive domain into another factorially familiar domain, in Cattell's case this was personality, it was thought possible to form a more accurate baseline for rotation. Using principal axis analysis, the scree test for significant factors, a simple structure oblique rotation followed by a 'blind' rotation technique devised by himself, Cattell obtained a g_f factor and what appeared to be Thurstone primary factors of verbal, numerical and reasoning skills (figure 5.4a). Second-order factor analysis, that is repeating the analysis using the factors from the first analysis (first order) as the

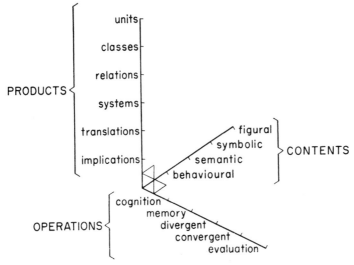

FIGURE 5.5. Guilford's model of the intellect

vectors, revealed g_f again and a second factor g_c (figure 5.4b) which combined the 'primaries' of the first-order analysis. This process was repeated for the third and fourth orders, the final order giving g_f and g_c in different factors, but with some correlation between them, as might have been expected from the use of oblique techniques.

Possibly the most ambitious contemporary application of factorial prescriptions appears in the work of Guilford.[8] He is most critical of the g theory[9] and in many respects his theory is diametrically the opposite of it. His researches led him to postulate three independent basic intellectual dimensions of *operations*, *content* and *products* each of which subdivide into categories (five operational, four intellectual content, six intellectual products). By arranging these in a three-dimensional array ($5 \times 4 \times 6$), 120 intellectual factors or abilities are generated, as in figure 5.5, of which Guilford now claims to have extracted and isolated about eighty.

Guilford's use of factorial solutions is a good illustration of hypothetico-deductive method. His model proliferates numerous testable factors which he proceeds to verify using test material presumed to fulfil the specifications derived from the categories along the dimensions. In a recent study,[10] for example, he took fifty-nine variables mostly from his model, applied principal axis analysis with estimates of the communality in the leading diagonals, took out twenty-five from forty-one extracted factors and submitted these for orthogonal rotation to simple structure. Factors exhibited clusterings of tests in keeping with the model.

Sufficient has been said to convince the reader of the wide application of factorial procedures in the study of human ability and the complexity of the problems with which the factorist wrestles. It is not really a simple case of right or wrong methods, but which are most representative, instructive and illuminating. Ultimately, as with any study aspiring to scientific status, the cogency of the theories will depend on their reliability in solving relevant problems.

PERSONALITY

Vernon (1964),[11] in his book on personality assessment, outlines three broad approaches to the interpretation of personality. These are 'naive', intuitive (psychoanalytic approaches such as Freud and his followers) and inferential studies. It is the latter approach which has relied extensively on psychometric evaluation and in which theorists have availed themselves of factorial solutions. Two prominent researchers in this field are Eysenck and Cattell whose work will form the substance of this section on personality.

One fertile and compelling point of view over the last thirty years is that of Eysenck.[12] During this period he has unfolded a theory of personality structure containing a limited number of fundamental dimensions in keeping with the principle of parsimony. In fact, he has reduced personality organization to three basic dimensions.

namely, extraversion–introversion, neuroticism and psychoticism. He also makes a case for a fourth dimension of intelligence. This atomistic approach to the study of personality (or perhaps molecular when compared with Cattell as we shall see) does not exclude the presence of any one dimension. On the contrary, an adequate description of an individual's personality would require a set of scores obtained from the three dimensions specified above. Since the dimensions are orthogonal (uncorrelated), the scores would enable us to represent that individual by using some point in three-dimensional space, thus enabling a description of his personality in terms relative to the population norms for the dimensions.

The construction of this framework of statistically independent elements of personality owes a lot to factorial design. An example of Eysenck's attempt to derive the dimension of neuroticism[13] will help to clarify the interesting methods he uses. An extensive collection of tests involving personality inventories and questionnaires (e.g. The Maudsley Medical Questionnaire, Minnesota Multiphasic Personality Inventory, social attitudes) and 'objective behaviour' tests (e.g. dexterity, persistence, speed of tapping, body sway, concentration) was administered to a sample from the population (assumed 'normal') and a 'neurotic' group, the latter being distinguished by psychiatric symptoms (patients exhibiting hysteric and anxiety states). Product-moment correlations between the test scores for the normal group, followed by a centroid analysis, gave a first common factor (extracting about 20 per cent of the total variance for sixteen tests) tentatively labelled a factor of neuroticism.

The next task was to devise a criterion against which the validity of the first factor could be tested. This procedure, created by Eysenck, is known as *criterion analysis* though Cattell prefers *criterion rotation* as a more appropriate label. The first step was to calculate the correlation between the test scores and the dichotomy of 'normal' or 'neurotic'. A special kind of correlation known as biserial correlation* is required for this. If most of the neurotics obtained a high score on a test purporting to measure neuroticism and the normals obtained low scores,

---

*A biserial correlation is adopted when one of the variables to be correlated is a dichotomy (boy/girl, young/old, normal/neurotic) and the other is a continuous scale (scores on a test of intelligence or personality questionnaires).

a high correlation would be expected by this method. The tests are being correlated for their ability to discriminate between normal and neurotics as defined by psychiatrists.

Thurstone's criteria have usually been the guide for rotation to

TABLE 5.1

Results from Eysenck's research

Test	C_N	F_1	D
Manual dexterity	·57	·405	·410
Motor control	·54	·644	·675
Non-suggestibility	·51	·620	·650
Persistence test I	·46	·607	·576
Personal tempo	·30	·438	·424
Dark adaptation	·27	·392	·407
Speed test (1)	·27	·523	·515
Persistence test B	·26	·632	·599
Stress test	·24	·294	·300
Maudsley Medical Inventory	·23	·143	·127
Non-perseveration	·21	·207	·189
Speed test (2)	·17	·565	·529
Judgement discrepancy score	·10	·497	·529
Goal discrepancy score	·06	·100	·094
Index of flexibility	·05	·275	·303
Fluency	·03	·300	·299
Correlation between C_N–F_1	0·566		
Correlation between C_N–D		0·575	

Reproduced with permission from H. J. Eysenck, *The Scientific Study of Personality*, Routledge and Kegan Paul, London, 1952, p. 74.

simple structure. On the other hand, Eysenck argues that we could apply hypothetico-deductive method to rotation by devising a clearly defined criterion and comparing it with factors extracted. In the example above, the biserial correlations would be compared with the neuroticism factor obtained from the correlations of test scores for either the normal or the neurotic group. Rotation of the extracted factors can then be applied until the neuroticism factor and the criterion factor are maximally correlated.

An illustration reproduced from Eysenck's (1952, p. 74) original analysis should help to clarify the method.

The first column in table 5.1 headed C_N (criterion column) gives the biserial correlations between the test scores and the neurotic/ normal dichotomy. The next column, F_1, is the first factor extracted and identified as the 'neuroticism' factor. D column gives the loadings derived after criterion rotation to a maximum correlation between C_N and F_1 (with the test vectors as unity). In the example, the angle through which F_1 was rotated came to about 5°, and improved the correlation between C_N and F_1 very slightly from 0·566 to 0·575.

The angle required to get maximum correlation between F_1 and C_N is very small; as all the tests were chosen as measures of neuroticism, their average (the first centroid factor) is therefore meaningful and close to the optimum position. This optimum position is unique and invariant; the addition of new tests, or the subtraction of old ones, would not alter the position of D. The relatively high, positive correlation between C_N and F_1 is in line with our hypothesis that the factor is one of neuroticism and supports the view that neuroticism is a continuum. (p. 74)

By comparable procedures, Eysenck has derived the other two dimensions suggested earlier. Nevertheless, these are early days in the construction of reliable personality assessments. There is still some degree of inconsistency in the supposedly most suitable tests for a given dimension and in the power of each test to discriminate the mentally deviant from 'normal' individuals.

Cattell[14] in the United States has also made a major contribution to the study of personality structure. He distinguishes between *source traits* which are underlying sources of observed behaviour and combinations of these traits which give *surface traits* (*syndromes* if they are abnormal). Neuroticism, defined in Eysenckian terms, would be a surface trait, although Eysenck prefers to refer to his dimensions as personality *types*. The cluster of behaviours characterizing neuroticism (shyness and withdrawal for example) he refers to simply as *traits*. Figure 5.6 summarizes the parallel between the terminology of these researchers.

Cattell accepts some overlap between the source traits and,

after a principal axis analysis of questionnaire responses covering the source traits,* he proceeds to an oblique rather than an orthogonal rotation. This has given sixteen personality factors at the source trait level (reproduced in Cattell's 16 PF test). Since these factors are oblique, some correlation occurs between them and consequently it is possible to resolve them into a higher-order structure. When this is done a close similarity occurs between these higher-order factors and Eysenck's dimensions of extraversion and neuroticism which appear to be more replicable than the 16 personality factors of Cattell (Eysenck and Eysenck, 1969).

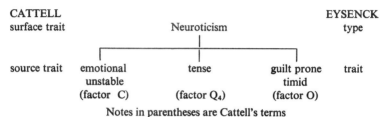

FIGURE 5.6. The relationship between the terminology of Cattell and Eysenck

There is clearly a great deal more to these studies than this superficial summary has portrayed. Both Eysenck (e.g. 1969) and Cattell have produced an enormous volume of research on personality structure of which the reader has been given a glimpse at some of the factorial designs employed.

OSGOOD'S SEMANTIC DIFFERENTIAL

In their book *The Measurement of Meaning*, Osgood *et al.*[15] seek to devise an objective measure of the meaning which people ascribe to various concepts relating to objects, ideas or people. The precise method will be demonstrated later, but for the moment, the respondent

---

*In the first place the source traits are hypothesized from observed behaviour ratings and confirmed from factor patterns. Questions which are thought to be potent in discriminating the source trait are then chosen and re-analysed for comparison with the original patterns obtained.

is presented with a concept and a large number of 'bipolar' adjectives by which he can register his feelings about the concept. The bipolar adjectives consist of such opposites as good–bad, strong–weak, hard–soft and so forth.

Fundamental to Osgood's theory is that associated with a given concept is a 'semantic space' which is derived as the outcome of an individual's experience and enables him to express his feelings about the concept in verbal form. Around each concept is

> ... a region of some unknown dimensionality and Euclidean in character. Each semantic scale, defined by a pair of polar (opposite-in-meaning) adjectives, is assumed to represent a straight line function that passes through the origin of this space, and a sample of such scales then represents a multi-dimensional space. The larger or more representative the sample, the better defined is the space as a whole. (p. 25)

Essentially, we are asked to imagine a concept at the centre of a semantic space with bipolar adjectives bristling out in all directions as the starting point for defining that concept.

The scales used in the semantic space can be divided into any convenient number of intermediate subdivisions between the extreme polar adjectives. These divisions would represent shades of favourable or unfavourable attitude as expressed in the adjectives selected. The number of divisions on a scale is usually seven and always an odd number so that the central division can represent 'neutral' feeling; in other words, if the respondent always marked the centre of each bipolar scale in response to a concept it would represent neutral feeling toward that concept. The concept and scales are usually presented to the subject in the following form:

CONCEPT

HAPPY :...:....: X .:....:....:....:....: SAD

FAST   :...:....:....:....:....:.X.:....: SLOW

BAD   :...:....:....: X .:....:....:....: GOOD   etc.

Instructions are given as to the shades of meaning to be associated with different points along the scale. Judgements are made by the subject by placing a cross at a point along the scale where it is most appropriate for him. In this way, the subject is able to indicate his feelings about the concept in terms of the scales.

> By semantic differentiation, then, we mean the successive allocation of a concept to a point in the multi-dimensional semantic space by selection from among a set of given scaled semantic alternatives. Differences in the meaning between two concepts is then merely a function of the differences in their respective allocations within the same space . . . (Osgood *et al.*, p. 26)

The centre of the scales, which might be imagined as criss-crossing the semantic space, could be taken as the origin of the semantic space; thus one can imagine the space being filled out by scales which intersect at their centres. The direction and distance of the subject's point from the origin could be used as an index of the quality and intensity of meaning which the subject places on the concept.

Osgood submitted a large number of scale scores to factor analysis using Thurstone's centroid method and orthogonal rotation. This was done using the intercorrelations between an individual's response on one scale with his responses on every other scale. This would furnish him with evidence for consistencies in individual responses. The first three factors accounted for a large portion of the total variance. These dominant factors were named the *evaluative*, the *potency* and the *activity* dimensions. Scale adjectives typical of the evaluative factor are good–bad and complete–incomplete. Typical potency adjectives are hard–soft, strong–weak and some activity adjectives are active–passive and hot–cold.

The first factor to emerge, the evaluative factor, accounted for between half to three-quarters of the total variance. It has been extensively employed as an attitudinal variable and correlates quite well with other attitude scales.

AN EXAMPLE FROM EXPERIMENTAL PSYCHOLOGY

The work of Fleishman and Hempel[16] provides a useful demonstration of factor patterns being used for the purposes of describing changes in the performance of a task. Briefly, the task, a complex coordination exercise used for testing pilot aptitude, required the subjects to respond to a pattern of visual signals by adjusting an aircraft control stick (hands) and a rudder (feet) simultaneously. The task was repeated with different patterns over a given period of time to assess the improvement in performance. Subjects completed eight consecutive practice sessions each consisting of five two-minute trials. In addition,

TABLE 5.2

Diagrammatic reproduction of Fleishman and Hempel's factor matrix

Variable		Factor						
		I	II	III	IV	V	VI	VII
Stages of practice in psychomotor task	1	☐	(✓)	☐	(✓)		(✓)	(✓)
	2	[✓]	(✓)	[✓]	○	(✓)	○	
	3	[✓]	(✓)	[✓]	(✓)	(✓)	○	☐
	4	✓	[✓]	✓		(✓)	☐	(✓)
	5	(✓)	[✓]	(✓)	☐	☐		○
	6	(✓)	[✓]	(✓)	☐	☐	☐	
	7	✓	✓	(✓)		☐		☐
	8	(✓)	✓	✓	☐		☐	☐
tests of coordination	9	✓						
tests of rate of movement	10		✓					
tests of spatial relations	11			✓				
tests of perceptual speed	12				✓			
tests of visualization	13					✓		
tests of mechanical experience	14							✓

✓ represents significant loadings
circles represent three highest loadings
squares represent three lowest loadings

they were invited to complete a battery of pencil and paper tests, and other apparatus, to determine psychomotor coordination, rate of movement, spatial relations, perceptual speed, visualization (ability to make mental manipulations of visual images), and mechanical experience.

Product-moment correlations, a centroid analysis and orthogonal rotation to simple structure gave a factor pattern which was used to discover the importance of motor and non-motor skills at different stages of practice in a psychomotor task. A most ingenious use was made of the loading sizes of the eight practice stages in relation to the remaining variables (see table 5.2).

The highest loadings for the earlier stages of practice occur in factors II, IV, V, VI and to some extent in VII. The researchers concluded that psychomotor coordination (II), spatial relations (IV), perceptual speed (V), visualization (VI) and to a lesser extent mechanical experience (V) played a major role in the early stages of practice. These factors are, in the main, 'non-motor' and it would seem that they become less and less important in individual performance as the practice period continues.

At the same time, the 'motor' factors of psychomotor coordination (II) and rate of movement (III) become more important with continued practice because the significant loadings appear later in the practice stages. This is also true of the first factor (I) which is clearly specific to the psychomotor task and in which the most significant loadings appear towards the end of the practice session. Apart from the second factor, which is hybrid in having loadings which are almost identical, the circles and squares will indicate at a glance the stages in the practice session, early or late, which contribute most variance in a particular factor.

THE DERIVATION OF LOCAL AND COSMOPOLITAN SCALES

Merton,[17] an American sociologist, has recently maintained that the orientations of people to their employing organizations can be described broadly as 'local' or 'cosmopolitan'. Locals (or organizationals) tend to be loyal to their employers; they seek recognition and promotion within their work institutions and identify themselves with its objectives. Conversely, cosmopolitans (or professionals) look to the wider world for personal satisfaction; they seek status recognition from professional and specialist organizations outside their work institutions; they are not particularly loyal to their employers, would be more likely to move around in their employments and look for approval from respected peers.

The first attempt to devise a questionnaire capable of distinguishing between locals and cosmopolitans appears in the work of Goldberg *et al.*[18] in the United States. In this country, the author[19] attempted a similar analysis using a sample of university students. Ten 'professional' and ten 'organizational' criteria were selected from Goldberg's questionnaire on the basis of their apparent relevance to English working conditions and English university students. A random sample of 100 student responses for the questionnaire were intercorrelated, subjected to Principal Components analysis and Varimax rotation with estimated communalities and five factors extracted and rotated using Kaiser's criterion. These factors accounted for 62 per cent of the total variance.

The first two factors accounted for 55 per cent of the total variance and only these contained the recognizable dimensions of 'local' and 'cosmopolitan' in separate factors. The existence of separate dimensions rather than a continuum of local/cosmopolitan tendencies was in line with Goldberg's findings in America. To derive final scales of local and cosmopolitan orientation, six items from factor I with loadings beyond the one per cent level of significance and a similar number from factor II were chosen. No item selected from one scale had significant loading on the alternative factor. These items were then used as the basis for a local/cosmopolitan measure.

Instructions for completing the items were as follows:

> When you are established in your career you will come up with scientific or technical ideas and will have to weigh up their value. You will have to make up your mind whether it is worth pressing to have them put into effect. In deciding whether you would press on with an idea, what weight do you think you would attach to the following considerations? Put 5 at the side of the consideration you think of the utmost importance, which you would always take into account; 4 if you think it would be usually important; 3 of occasional importance; 2 rather important; 1 not important at all.

The scales employed and loadings associated with each item were:

Scale of local (organizational) orientation

1. It would improve my firm's (or organization's) standing in the eyes of its customers or clients.   (·660)
2. It would improve the quality of the firm's (or organization's) product.   (·664)

3. It would help to achieve the firm's (or organization's) goals. (·566)
4. It is in line with the preferences of my managers and superiors. (·545)
5. It would increase my firm's profits. (·499)
6. It would further the company's (or organization's) growth. (·424)

Scale of cosmopolitan (professional) orientation

1. I should enjoy working on the idea. (·708)
2. The special interest I have in the particular scientific/ technical field. (·678)
3. It would give me the opportunity to do good research. (·502)
4. It would keep people who work under me happy. (·394)
5. It is original and creative. (·312)
6. It has a theoretical relevance to existing knowledge in the field. (·328)

Clearly, the two sets of items closely resemble the definitions proposed above where the 'local' is more concerned with promoting the welfare of his employing institution whilst the 'cosmopolitan' looks for personal gratification and prestige.

MEDICINE AND PUBLIC HEALTH

We conclude this chapter with three rather interesting examples selected from medical and public health research. In all cases a principal axis method was used followed, in some cases, by rotation. These examples are included to give some idea of the kinds of problem being tackled by factor analysts in medical fields.

Cady[20] and his colleagues were interested in finding the personal, familial, physical and biological characteristics typical of coronary artery patients. They administered Cattell's 16PF adult personality test, Sheldon's somatotype criteria for body shape, and made various medical tests of cholesterol and blood pressure using a sample of 61 patients and 146 controls chosen at random. The prominent factor they uncovered, which related some personality factors with cholesterol content in the blood, diastolic blood pressure and coronary symptoms, substantiated the subjective assessments of doctors and subsequently brought an element of certainty into diagnosis where doubt had existed previously. The findings were also suggestive of several additional investigations particularly in relating personality traits to various diseases.

An attempt to introduce more objective evaluation into therapeutic effectiveness was the primary source of motivation in the researches of Petrinovich and Hardyck.[21] The practitioner has often to cope with a multitude of symptoms in varying stages of progress and from these he must make some judgement as to the nature and extent of the condition and, as in the present study, to arrive at an early decision about the effectiveness of therapy. Sound motion pictures of performance in a motor task were recorded for 40 patients with Parkinson's disease before and after an operation to alleviate uncontrolled motor activity. Performance was rated by a panel of experts using an established list of motor activities characteristic of Parkinson patients. Performance was judged before, three months after and at intervals of one year after the operation. The successive factor analyses of these judgements showed a shift in emphasis from gross to fine motor behaviour in pre- and post-operational trials. Much more important than this rather obvious conclusion was the appearance of three additional factors in the post-operational data, namely, fine motor skills, tremor and finger movement, which could serve as yardsticks for ascertaining the rate of progress of patients. The authors conclude that:

> Advanced statistical techniques such as factor analysis allow for a precise and efficient solution to the complex problem of evaluating the multivariate changes in behaviour that frequently occur as a result of neurosurgical procedures.

As a final illustration of factor applications, we shall look at the work of Jenkins and Zyzanski[22] in public health on the public's reaction to various serious diseases as judged by their beliefs and feelings. A large urban community (436 adults between 20 and 39 years of age) were presented with a number of scales relating to perceptions and feelings about poliomyelitis, cancer and mental illness. The questions endeavoured to disclose people's thoughts about public and personal susceptibility to the diseases, risk of death or disability, whether the disease is thought of as clean or dirty, proud or disgraceful—including a moral judgement about the kind of people attacked by the disease. Several intriguing factors emerged of which three will be mentioned. A 'Human-Mastery' factor (the labels belong to the researchers) included items about the extent to which human intervention was effective. The researchers suggested that this dimension was useful in predicting and influencing public response to health programmes. For diseases which the public knew were not under control, as in the case of cancer, there appeared to be less likelihood of their coming forward for preventative medicine. This, apparently, occurred when polio vaccine first came on the market. The public were not at all keen to take advantage.

A second factor concerned 'social-acceptable or social-stigma' responses particularly with respect to polio and cancer. The sample was conscious of some diseases bringing with them 'an aura of prestige' whilst others were an embarrassment. The image of famous people who succeeded whilst being afflicted with polio paralysis (Roosevelt is an obvious case) was thought to be the important influence here.

A third group of items reflected feelings of personal involvement where people often talked or thought about a disease and weighed up their chances of getting it. When members of the public felt personally involved there would naturally be a greater readiness for them to respond to invitations for preventative treatment. The conclusion reached by these investigators was to endeavour, in the case of diseases for which people would not come forward for treatment, to whip up enthusiasm and feelings of personal involvement and public duty. This was, in fact, the method adopted in the polio vaccination campaigns during the early years when it became clear that the public were not coming forward.

Factor analysis has clearly made an impressive contribution in a variety of important practical ways. Equally, it has still much to offer in the exploration of multivariate enquiries. With all its limitations and imperfections, it has proved to be one of the most serviceable and productive tools at the disposal of behavioural scientists. With this in mind, it is hoped that this slim volume has provided, as far as possible, answers to some of the less technical questions about the technique for those who need to refer to literature containing factor analytical solutions.

REFERENCES

1. For a splendid up-to-date discussion of the subject, have a look at H. J. Butcher, *Human Intelligence: Its Nature and Assessment*, Methuen, London, 1968.

2. C. Spearman, 'General intelligence, objectively determined and measured', *Am J. Psychol.*, **15**, 201–293 (1904). As suggested in Chapter 1, he drew on the two types of ability postulated by Galton.

3. For a thorough treatment of the Two-Factor Theory see C. Burt, 'The two-factor theory', *Br. J. Psychol.: Statist. Section*, **2**, 151–179 (1949). R. B. Cattell, *Factor Analysis*, Harper, New York, 1952 also takes up this issue in chapter 4. In both these references Spearman's use of corrections for correlation coefficients, 'tetrad difference' technique and 'hierarchical pattern' are said to be responsible for this pattern appearing.

4. C. Burt, *The Distribution and Relations of Educational Abilities*, King, London, 1917. See also Thomson's work, for example, W. Brown and G. H. Thomson, *The Essentials of Mental Measurement*, Cambridge University Press, Cambridge, 1921.

5. The method is discussed in K. J. Holzinger and H. H. Harman, *Factor Analysis*, University of Chicago Press, Chicago, 1941.

6. P. E. Vernon gives some useful details of this theory in *The Structure of Human Abilities*, 2nd ed., Methuen, London, 1961.

7. R. B. Cattell, 'Theory of fluid and crystallized intelligence: a critical experiment', *J. educ. Psychol.*, **54**, 1–22 (1963); 'The theory of fluid and crystallized intelligence checked at the 5–6 year-old level', *Br. J. educ. Psychol.*, **37**, 209–224 (1967). This latter paper also gives a useful demonstration of various factorial techniques.

8. Many papers and monographs have appeared between 1960 and the present, published by the University of Southern California. Some of the important aspects of Guilford's thinking are summarized in 'Intelligence: 1965 Model', *Am. Psychol.*, **21**, 20–26 (1966).

9. For example, J. P. Guilford, 'The structure of intelligence' in *Handbook of Measurement and Assessment in the Behavioral Sciences* (Ed. D. K. Whitla), Addison-Wesley, Reading, Mass., 1968.

10. R. Hoepfner and J. P. Guilford, 'Figural, symbolic, and semantic factors of creative potential in ninth-grade students', *Report No. 35, Psychological Laboratory*, University of Southern California, 1965.

11. P. E. Vernon, *Personality Assessment: a Critical Survey*, Methuen, London, 1964.

12. H. J. Eysenck, *The Dimensions of Personality*, Routledge and Kegan Paul, London, 1947; *The Scientific Study of Personality*, Routledge and Kegan Paul, London, 1952 are amongst the earliest of his expositions. Chapter 8 in the latter gives a summary of his views on the organization of personality. See also H. J. Eysenck and S. B. G. Eysenck, *Personality Structure and Measurement*, Routledge and Kegan Paul, London, 1969.

13. Eysenck (1952) note 12 above.

14. J. B. Cattell, *Personality and Motivation Structure and Measurement*, Harrap, London, 1957; *The Scientific Analysis of Personality*, Penguin, London, 1965.

15. C. E. Osgood, G. J. Suci, and P. H. Tannenbaum, *The Measurement of Meaning*, University of Illinois Press, Urbana, 1957.

16. E. A. Fleishman and W. E. Hempel, 'Changes in factor structure of a complex psychomotor test as a function of practice', *Psychometrika*, **19**, 239–252 (1954).

17. R. K. Merton, *Social Theory and Social Structure*, Free Press, Glencoe, Ill., 1964.

18. L. C. Goldberg, F. Baker and A. H. Rubenstein, 'Local-cosmopolitan: unidimensional or multidimensional?' *Am. J. Sociol.*, **70**, 704–710 (1965).

19. D. Child and F. Musgrove, 'Career orientations of some university freshmen', *Educ. Rev.*, **21**, 209–217 (1969).

20. L. D. Cady, M. M. Gertler, L. G. Gottsch and M. A. Woodbury, 'The factor structure of variables concerned with coronary artery disease', *Behavl Sci.*, **6**, 37–41 (1961).

21. L. Petrinovitch and C. Hardyck, 'Behavioural changes in Parkinson patients following surgery: a factor analytic study', *J. Chron. Dis.*, **17**, 225–233 (1964).

22. C. D. Jenkins and S. J. Zyzanski, 'Dimensions of belief and feeling concerning three diseases, poliomyelitis, cancer and mental disease: a factor analytic study', *Behavl Sci.*, **13**, 372–381 (1968).

FINDING FACTOR LOADINGS USING SIMPLE SUMMATION METHOD

As an example of a procedure for calculating the factor loadings, for the mathematically inclined, we shall look at a method devised by Burt. The illustration in figure 2.7 in which five tests were analysed will be used again in this simple summation method to show that the calculations lead to the same results. The numerical procedure will be described first, followed by a geometrical proof.

The first task is to draw up the complete correlation matrix

TABLE A.1

	T_1	T_2	T_3	T_4	T_5
T_1	1·0000	0·9848	0·3420	0·0000	−0·1736
T_2	0·9848	1·0000	0·5000	0·1736	0·0000
T_3	0·3420	0·5000	1·0000	0·9397	0·8660
T_4	0·0000	0·1736	0·9397	1·0000	0·9848
T_5	−0·1736	0·0000	0·8660	0·9848	1·0000
sum of columns	2·1532	2·6584	3·6477	3·0981	2·6772 = 14·2346
loadings	0·572	0·705	0·970	0·824	0·711

$$\sqrt{14\cdot2346} = 3\cdot773$$

including the values for both upper and lower triangles with the leading diagonal containing either unity or the communality estimate. As our original illustration employed unities we shall continue to use them in these calculations. Table A.1 above is the correlation matrix from figure 2.7 drawn up in full.

Next, each column is totalled and the sum of all the five column totals determined. In this case it is 14·2346. The square root of this quantity is found and gives 3·773. Finally, the value of the square root is divided into each column total. For example, in the first column T_1, the total was 2·1532 and this divided by 3·773 gives 0·572. The result obtained is the loading of test 1 on factor I. Continuing to divide each column total by the square root of the grand total will give the remaining loadings seen recorded at the bottom of each column. The values for this first factor, allowing for rounding errors,

TABLE A.2

Vector diagram and matrix for a general case with five factors

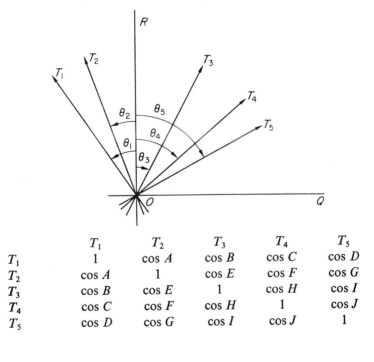

	T_1	T_2	T_3	T_4	T_5
T_1	1	cos A	cos B	cos C	cos D
T_2	cos A	1	cos E	cos F	cos G
T_3	cos B	cos E	1	cos H	cos I
T_4	cos C	cos F	cos H	1	cos J
T_5	cos D	cos G	cos I	cos J	1

come very close to those displayed in the factor matrix of table 2.5.

For justification of the process described above we can resort to a geometrical proof. Table A.2 shows a general case for five test vectors all, fortuitously, bearing correlations which enable an arrangement in the plane of the paper.

Let the angles *between the tests* be represented by the letters of the alphabet as shown in the matrix in table A.2. Thus angles *A* to *J* represent all the possible angles between the test vectors 1 to 5. In the vector diagram for these tests, the angles *between the test vector and the resultant OR* have been given the values θ_1 to θ_5. Of course, angle A is equal to $\theta_1 - \theta_2$, and *E* is equal to $\theta_2 + \theta_3$ and so on through the table, but for the purposes of deriving the relationships we need, it will be more convenient to express the angles in two ways.

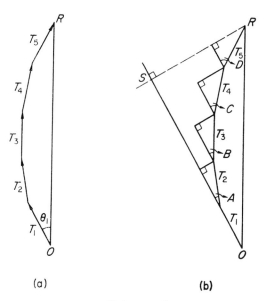

(a)   (b)

FIGURE A.1. Polygon of vectors

One arrangement for finding the resultant OR is to build up a polygon of vectors by placing the vectors 'head to tail' in the order T_1 to T_5 with the precise lengths (T_1, T_2, T_3, T_4 and T_5) and angles taken into account. The lengths are all the same (unity in fact, although for the moment we shall argue the general case) so the polygon will look something like the figure A.1(a) above. By joining the 'tail' of T_1 to the 'head' of T_5 we obtain the magnitude and direction of the resultant OR [OR is equal in value to the latent root (eigenvalue) referred to in chapter 3].

First, extend vector T_1 in the direction OS and drop a perpendicular from point R onto the extended vector to meet it at S.
Then

$$\cos \theta_1 = \frac{OS}{OR}$$

and

$$OR \cos \theta_1 = OS \tag{1}$$

But OS can be expressed in terms of the other vectors and the angles existing between them. Figure A.1(b) illustrates this point that by forming triangles for each vector with one side parallel to OS and the other parallel to OR we create a series of triangles with the angles A to D as shown. The length OS can now be expressed as

$$OS = T_1 + T_2 \cos A + T_3 \cos B + T_4 \cos C + T_5 \cos D$$

Substituting from equation (1) and replacing T_1 to T_5 by 1 (because the vectors are made equal to unity) we obtain

$$OR \cos \theta_1 = 1 + \cos A + \cos B + \cos C + \cos D \tag{2}$$

Using the second vector T_2 for the same purpose, we obtain

$$OR \cos \theta_2 = \cos A + 1 + \cos E + \cos F + \cos G \tag{3}$$

Similarly for the remaining vectors we have

$$OR \cos \theta_3 = \cos B + \cos E + \quad 1 \quad + \cos H + \cos I \tag{4}$$

$$OR \cos \theta_4 = \cos C + \cos F + \cos H + \quad 1 \quad + \cos J \tag{5}$$

$$OR \cos \theta_5 = \cos D + \cos G + \cos I + \cos J + \quad 1 \tag{6}$$

But the right-hand sides of equations (2) to (6) give the entire matrix of table A.2. Therefore, if we add the left-hand sides of these equations it gives

$$OR(\cos \theta_1 + \cos \theta_2 + \cos \theta_3 + \cos \theta_4 + \cos \theta_5) = \text{Sum of all the matrix} \tag{7}$$

It can be shown from figure A.1(b) that

$$\cos \theta_1 + \cos \theta_2 + \cos \theta_3 + \cos \theta_4 + \cos \theta_5 = OR$$

Consequently, equation (7) becomes

and

$$OR^2 = \text{Sum of all the matrix}$$

$$OR = \sqrt{\text{Sum of all the matrix}}$$

Reference to table A.1 will reveal that the sum of the matrix was 14·2346.

The loading for T_1, it will be recalled, is the cosine of the angle between the test vector and the resultant which in this case is $\cos \theta_1$. From equation (2)

$$\cos \theta_1 = \frac{1 + \cos A + \cos B + \cos C + \cos D}{OR}$$

But the numerator is, in fact, the sum of all the values in the first column of the matrix in table A.2. Also OR is the square root of the sum of the entire matrix.

$$\therefore \text{ the first loading} = \frac{\text{sum of all the values in the first column}}{\text{the square root of the sum of the entire matrix}}$$

which is exactly the procedure adopted.

Finding the second factor loadings is rather more tedious. To begin with, the vector directions need some adjustment to give the closest possible clusters of the vectors. A resultant along OQ at right-angles to OR as it stands in table A.2 is not a particularly satisfactory set-up. However, by changing the signs and thus reversing the direction of T_1 and T_2 a better cluster is created without affecting the mathematical accuracy of the result. We have also to allow for that part of the common variance accounted for by the first factor. To do this we must look more closely at the link between the factor loadings and the original correlation matrix.

Let the loadings for tests 1 to 5 be l_1, l_2, l_3, l_4 and l_5 respectively. The first entry of unity in the matrix of table A.1 is the self-correlation for test 1. The loading for this test is l_1, consequently the 'self-correlation' accounted for by the first factor will be l_1^2. The entry in table A.1 has therefore been reduced to $1 - l_1^2$. For the numerical example in table A.1 it would be $1 - (0 \cdot 572)^2 = 1 - 0 \cdot 3272 = 0 \cdot 6728$. This figure is the residual correlation and constitutes the first entry for test 1 in a new matrix of *residues* which will become the starting point for the extraction of the second factor. Similarly, the entry for the correlation between T_1 and T_2 will be reduced by an amount equal to $l_1 l_2$. From table A.1 this is $0 \cdot 9848 - l_1 l_2 = 0 \cdot 9848 - (0 \cdot 572 \times 0 \cdot 705)$ which gives $0 \cdot 9848 - 0 \cdot 4033 = 0 \cdot 5815$ for the $T_1 T_2$ entry. Continuing in this fashion by subtracting the part explained by the first factor from the corresponding original matrix entry, a matrix of residues can be found from which the second factor is extracted. Subsequent factors can be discovered in exactly the same way. Readers might like to derive the second factor as an exercise—it should compare with the second factor in table 2.5. Moreover, there should not be a residual matrix because all the variance was accounted for by the first two factors.

For those interested in pursuing the matter further, Thomson, in his book *The Geometry of Mental Measurement*, gives an example of simple summation using estimates of the communalities in the leading diagonals.

SIGNIFICANCE LEVELS FOR PEARSON PRODUCT-MOMENT CORRELATION COEFFICIENTS

	Values of correlations required	
Sample size	at 5% level	at 1% level
5	·755	·875
10	·576	·714
15	·483	·605
20	·425	·538
25	·380	·488
30	·338	·440
35	·320	·417
40	·300	·394
45	·280	·370
50	·262	·346
60	·248	·328
70	·233	·308
80	·220	·290
90	·206	·272
100	·194	·255
150	·158	·209
200	·137	·182
250	·125	·163
500	·088	·115

Interpolated from graphical Table X in G. H. Fisher, *The New Form Statistical Tables*, University of London Press, London, 1965. Reproduced with the kind permission of the author.

SIGNIFICANCE LEVELS OF FACTOR LOADINGS USING THE BURT–BANKS FORMULA*

(When $N = 50$ and $N = 100$)

To give readers an idea of the changes in loading values ensuing when corrections are applied to the standard error of a correlation coefficient, two tables have been prepared for selected sample sizes of 50 and 100. The formula* is

$$\text{Standard error of a loading} = \text{Standard error of a correlation} \left(\sqrt{\frac{n}{n+1-r}} \right)$$

where $n =$ the number of variables in the analysis, and $r =$ the factor number, that is the position of the factor during extraction.

The standard error of a correlation can be obtained from the table in Appendix B.

Suppose 200 individuals provide 30 sets of test scores which are then subjected to factor analysis. What magnitude of loading would be required in order to reach significance in, say, the twentieth factor to be extracted?

The standard error for a sample of 200 at the one per cent level is 0·182 (see Appendix B). Also

$$n = 30$$

and

$$r = 20$$

---

*The formula is quoted in C. Burt and C. Banks, 'A factor analysis of body measurements for British adult males', *Ann. Eugenics*, **13**, 238–256 (1947).

Substituting in the equation above, the standard error of the loading would be

$$0{\cdot}182 \sqrt{\frac{30}{30 + 1 - 20}} = 0{\cdot}182 \sqrt{\frac{30}{11}} = 0{\cdot}300$$

In other words, for loadings to satisfy the one per cent level of significance in the twentieth factor they must be at least $\pm 0{\cdot}300$. With smaller samples, as shown in table C.1 and C.2, the value required becomes quite large.

TABLE C.1

$N = 50$

Number of variables	Significance level as percentage	Factor number											
		1	2	3	4	5	6	7	8	9	10	15	20
10	5	·262	·276	·292	·312	·337	·370	·414	·478	·585	·828	—	—
	1	·346	·364	·386	·413	·446	·489	·547	·631	·773	—	—	—
20	5	·262	·268	·276	·284	·292	·302	·312	·324	·337	·353	·478	—
	1	·346	·354	·364	·375	·386	·399	·413	·429	·446	·466	·631	—
30	5	·262	·266	·270	·276	·281	·286	·292	·298	·305	·312	·358	·432
	1	·346	·351	·357	·364	·371	·378	·386	·394	·403	·413	·473	·571
40	5	·262	·265	·268	·272	·276	·279	·284	·288	·292	·297	·324	·361
	1	·346	·350	·354	·359	·364	·369	·375	·380	·386	·392	·429	·477
50	5	·262	·263	·264	·265	·267	·268	·270	·271	·273	·274	·282	·290
	1	·346	·349	·352	·356	·360	·364	·368	·372	·377	·381	·407	·439

TABLE C.2

$N = 100$

Number of variables	Significance level as percentage	Factor number											
		1	2	3	4	5	6	7	8	9	10	15	20
10	5	·194	·204	·216	·231	·250	·274	·306	·354	·433	·613	—	—
	1	·255	·268	·285	·304	·328	·360	·403	·465	·570	·806	—	—
20	5	·194	·198	·204	·210	·216	·223	·231	·240	·250	·261	·354	·867
	1	·255	·261	·268	·276	·285	·294	·304	·316	·328	·343	·465	—
30	5	·194	·197	·200	·204	·208	·212	·216	·221	·226	·231	·265	·320
	1	·255	·259	·263	·268	·273	·279	·285	·290	·297	·304	·349	·421
40	5	·194	·196	·198	·201	·204	·207	·210	·213	·216	·220	·240	·267
	1	·255	·258	·261	·264	·268	·272	·276	·280	·285	·289	·316	·351
50	5	·194	·195	·197	·200	·202	·204	·206	·208	·211	·214	·228	·246
	1	·255	·257	·260	·262	·265	·268	·271	·274	·277	·281	·300	·323

SUGGESTED FURTHER READING

C. J. Adcock, *Factor Analysis for Non-Mathematicians*, Melbourne University Press, Melbourne, 1955.

C. Burt, *The Factors of the Mind*, University of London Press, London, 1940.

C. Burt, 'Tests of significance in factor studies', *Br. J. Psychol.: Statist. Section*, **5**, 109–133 (1952).

R. B. Cattell, *Factor Analysis*, Harper, New York, 1952.

H. J. Eysenck, 'The logical basis of factor analysis', *Am. Psychol.*, **8**, 105–114 (1953).

B. Fruchter, *Introduction to Factor Analysis*, Van Nostrand, New York, 1954.

J. P. Guilford, 'When not to factor analyze', *Psychol. Bull.*, **49**, 26–37 (1952).

H. H. Harman, *Modern Factor Analysis*, University of Chicago Press, Chicago, 1967.

S. Henrysson, *Factor Analysis in the Behavioural Sciences*, Almqvist and Wiksell, Uppsala, 1957.

Q. McNemar, 'The factors in factoring behaviour', *Psychometrika*, **16**, 353–359 (1951).

E. A. Peel, 'Factor analysis as a psychological technique', *Symp. Psychological Factor Analysis*, Uppsala, 1953.

C. Spearman, 'What the theory of factors is not', *J. educ. Psychol.*, **22**, 112–117 (1931).

G. H. Thomson, *Factorial Analysis of Human Ability*, University of London Press, London, 1948.

G. H. Thomson, *The Geometry of Mental Measurement*, University of London Press, London, 1954.

L. L. Thurstone, *Multiple Factor Analysis*, University of Chicago Press, Chicago, 1947.

P. E. Vernon, *The Structure of Human Abilities*, 2nd ed., Methuen, London, 1961.

D. K. Whitla, (Ed.), *Handbook of Measurement and Assessment in the Behavioral Sciences*, Addison-Wesley, Reading, Mass., 1968, especially H. H. Harman, 'Factor analysis', pp. 143–170.

INDEX

A

Activity dimension (Osgood) 80
Adcock, C. J. 29, 101
AH5 39, 49 *reference* 2
Analytical rotation 53, 59–61
Angle between two vectors—*see* Correlation
Applications of factor analysis 5, 6, 67–86

B

Baker, F. 83, 88
Banks, C. 45–46, 97n
Bi-factor solutions 52, 70
Binormamin 61
Bipolar adjectives 79
Bipolar factors 48–49, 52
Biquartimin 61
Biserial correlation 75n, 76–77
Brebner, A. 41, 50
Brown, W. 87
Burt, C. 1, 5, 8, 9, 12, 45–46, 50, 52, 69, 70, 87, 97n, 101
Burt–Banks formula 46, 97
Butcher, H. J. 49, 87

C

Cady, L. D. 84, 88
Carroll, J. B. 61, 65
Cattell, R. B. 17, 29, 43, 44, 50, 60, 61, 65, 72–73, 75, 77–78, 87, 88, 101
Causal relationships 4, 9, 10, 14n
Centroid method 5, 23, 26–28, 52—*see also* Simple summation
Child, D. 38, 50, 83, 88
Christie, T. 49
Common factors 33, 69
Common factor vectors 23
Common variance—*see* Variance
Communality 35–36, 36n, 37, 41, 59, 62
Component analysis 36–38
Concomitant variation 1, 3, 4–5, 33
Conflict dimensions 9–10
Content (Guilford's model) 73–74
Convergent thinking tests 38n
Correlation, biserial 75n
concept of 3, 4, 14n
as cosine of angle between vectors 18–22
and factor loadings 23–24